The Grenfell Tower Fire

The Grenfell Tower Fire

A Firefighter's View

Tony Sullivan

WHITE OWL
AN IMPRINT OF PEN & SWORD BOOKS LTD.
YORKSHIRE - PHILADELPHIA

First published in Great Britain in 2025 by
White Owl
An imprint of Pen & Sword Books Limited
Yorkshire – Philadelphia

Copyright © Tony Sullivan 2025

ISBN 978 1 39906 446 0

The right of Tony Sullivan to be identified as
Author of this Work has been asserted by him in accordance
with the Copyright, Designs and Patents Act 1988.

A CIP catalogue record for this book is
available from the British Library.

All rights reserved. No part of this book may be reproduced,
transmitted, downloaded, decompiled or reverse engineered in
any form or by any means, electronic or mechanical including
photocopying, recording or by any information storage and retrieval
system, without permission from the Publisher in writing. No part of
this book may be used or reproduced in any manner for the purpose
of training artificial intelligence technologies or systems.

Typeset by Mac Style
Printed in the UK by CPI Group (UK) Ltd, Croydon, CR0 4YY.

The Publisher's authorised representative in the EU for product
safety is Authorised Rep Compliance Ltd., Ground Floor,
71 Lower Baggot Street, Dublin D02 P593, Ireland.
www.arccompliance.com

For a complete list of Pen & Sword titles please contact

PEN & SWORD BOOKS LIMITED
47 Church Street, Barnsley, South Yorkshire, S70 2AS, England
E-mail: enquiries@pen-and-sword.co.uk
Website: www.pen-and-sword.co.uk
or
PEN AND SWORD BOOKS
1950 Lawrence Road, Havertown, PA 19083, USA
E-mail: uspen-and-sword@casematepublishers.com
Website: www.penandswordbooks.com

Dedicated to the seventy-two people who lost their lives, the survivors and the firefighters who attended the Grenfell Tower fire on 14 June 2017

Contents

Tables		viii
Figures		ix
Acknowledgements		x
Introduction		xi
Part I	A Timeline and Career	1
Part II	From Lakanal to Grenfell	51
Part III	The Fire and Aftermath	97
Appendix: Timeline to Tragedy		151
Notes		166
Bibliography		171

Tables

Table 1: Fire safety legislation prior to 2005 — 79
Table 2: Timeline of refurbishment at Grenfell — 90
Table 3: First appliances despatched and arrival times — 103
Table 4: Pathways of fire spread — 110
Table 5: Initial assistance messages — 111
Table 6: Timeline 00:54-01:26 — 113
Table 7: Summary of timeline of fire — 126

Figures

Figure 1: High-rise procedure: Sectors　44
Figure 2: High-rise procedure　44
Figure 3: Grenfell Tower before refurbishment (*Wikimedia Commons*)　86
Figure 4: Cladding and insulation at Grenfell (*Wikimedia Commons*)　88
Figure 5: Flat 16 layout　101
Figure 6: Floor and flat numbers　106
Figure 7: Plan of fourth floor　107
Figure 8: Plan for floors 4-23　107
Figure 9: Grenfell Tower, 4:43 a.m. Wednesday, 14th June 2017 (*Wikimedia Commons*)　123

Acknowledgements

Many thanks to all the firefighters who assisted me and specifically to Steve Dudeney for allowing me to quote from his blog: *stevedude68: The life of a London Firefighter*.

Images

Figure 3: Grenfell Tower before refurbishment (Wikimedia Commons). CC BY-SA 2.0, https://commons.wikimedia.org/w/index.php?curid=59914872

Figure 4: Cladding and insulation at Grenfell (Wikimedia Commons). CC BY-SA 4.0, https://commons.wikimedia.org/w/index.php?curid=60129251

Figure 9: Grenfell Tower, 4:43 a.m. Wednesday, 14 June 2017 (Wikimedia Commons). CC BY 4.0, https://commons.wikimedia.org/w/index.php?curid=59913134

Introduction

On the morning of Wednesday 14 June 2017, I woke as normal around 6:30am. I was not due on shift until the next day and found myself alone in the kitchen in an unusually quiet house. My three young children were not yet up and the chaos of the morning school run had yet to begin. Like many across the capital that morning, I was totally unaware that, as I had slept safely in my bed, twenty-four miles across the other side of London, scores of other families had already suffered an unimaginable tragedy. I was still in the kitchen with a cup of tea when I got the first text. Steve Mewett, one of my crew managers at Addington Green Watch, asked me if I'd seen the news. 'What news?' I asked. He seemed lost for words. 'Just put the television on,' came the reply.

My other crew manager, Dave Taylor, texted soon after. More phone calls followed. None of us could quite believe what we were seeing. Sitting here writing this five years later some phrases come back to me: 'How could this happen?'. Someone used the phrase 'like Lakanal on steroids', a reference to the Lakanal House fire in 2009. In essence, Lakanal was an incident where neither the building nor the fire behaved as we'd expected it to. That, too, came as a complete shock to firefighters. But it was nothing like Grenfell. This was on another scale entirely. More phone calls with other colleagues followed and similar views of disbelief were expressed.

The nearest equivalent in terms of shock was watching 9/11 unfold although that was to prove a far greater tragedy in terms of lives needlessly lost. On that occasion I was on duty, visiting a school with the fire appliance when the first plane struck. Just as we were preparing to leave, a teacher came in with the news which we assumed at that point was some freak accident. The station was only a couple of minutes away from the school and in that short journey we naively wondered how a plane could accidentally fly into a building. It was only later I learned a plane had done just that in 1945, a B-25 bomber slamming into the Empire State Building in thick fog. We got back and turned the TV on just in time to see the second plane hit and tear away our assumptions along with several floors of the 110-storey Tower 2.

We sat there incredulous, and I remember that feeling of helplessness and confusion. Not quite believing it could be deliberate but knowing it couldn't now be an accident. Then the realisation that, along with what was undoubtedly going to be a huge death toll, there were bound to be significant losses of our fellow firefighters across the ocean. In the end 343 firefighters and paramedics were among the 2,753 who perished in the New York terrorist attacks. As the towers collapsed one after another the same question came to mind as we watched an entire block engulfed in flame sixteen years later: 'How could this be happening?'

A few days after the Grenfell fire a recording of a crew's reaction surfaced on social media. The crew had been called on in the early stages and it reveals their first impression as the incident came into view as they were en route. Some of the recording was inaudible but I've attempted to transcribe it accurately. It can also be seen on the BBC website dated 19 June 2017.[1] A fire appliance (what the public call a fire engine) has a crew of four to six firefighters and one of them must have had a mobile phone on and recorded the scene as they approached. As they turned a corner the entire face of the building could be seen ablaze. The shock in their voices is apparent as they realise this was not a block under construction or a derelict building:

> 'Jesus Christ mate'
> 'That is … is that … that's not a real block with people in it?'
> 'Fuck me.'
> 'Mate, how the fuck are we going to get into that …?'
> 'Whoa …'
> 'Fucking hell!'
> 'Jesus Christ!'
> 'Jesus!'
> (Inaudible swearing)
> 'Oh my God!'
> 'There's kids in there …'
> 'That's a real block!'
> 'Jesus'
> 'Mate, it's the Towering Inferno here?'
> 'How is that possible?'

'It's jumped up all the way along the flats, look!'
'How the fuck is that even possible?'
'How has that happened?'
'How is that even possible?'

How is that even possible? If there was ever a phrase to sum up the views of the rank and file of London's firefighters at the time of the fire, that is it. How is that even possible? Some readers who have followed the inquiry or know about the earlier Lakanal fire might find this reaction surprising. Whether they should have known is a different question and one that we will cover. What is important to remember is that this book is from the point of view of a firefighter and what I would like to make clear is that the size and nature of this incident came as a complete shock to many, if not all, at station level.

Why write this book and why now? One of the reasons is that many have been frustrated with how the fire has been portrayed in the press, and by politicians and commentators. At best often superficial and at worst disingenuous and misleading. The first days after the fire were awash with conspiracy theories which still linger today. However, the main reason is to try to describe accurately what went wrong and suggest ways to stop it happening again.

There's a concept called the 'Murray Gell-Mann Amnesia effect', which I believe comes from a speech in 2002 given by the writer Michael Crichton. Named after the American physicist Murray Gell-Mann, it posits that the media retains a level of credibility that is totally underserved even when we know articles touching on our own areas of expertise are partly or wholly inaccurate.

We are all experts in something, if not from our jobs or hobbies then our life experience. Reading an article on a subject we know a lot about can sometimes be a frustrating experience. One quickly gets the impression that it was written by someone with no knowledge of the subject at all. Yet when we turn the page onto something we are not experts on, such as international politics, a specific industrial dispute or a disaster, we are prone to forgetting just how poor the previous article was.

I've seen this myself in my career, whether it be an article regarding a small fire reported in a local newspaper or a major incident such as the *Marchioness* disaster. I've also had the misfortune to be involved in three periods of industrial action. The reporting came across to me as, more often than not, ranging from lazy incompetence to deliberate misinformation. This applies to Grenfell in spades. On subjects on which we are not expert, we are at the mercy of the press. Our very eagerness to learn makes us vulnerable to misinformation.

What can we do? Some will throw their hands up and declare it all 'fake news'. Others will search for some 'hidden truth' amongst the various conspiracy theory rabbit holes that proliferate online. But old habits die hard, and many will place their trust in whatever politician soothes their own prejudices and confirms their biases. In truth, the best place to find accurate information is with the experts. Someone with a few decades' experience in firefighting, fire safety or building control. I would ask the reader to think back on all the articles and interviews concerning Grenfell and consider how many included experts and how many were simply politicians or journalists? At the very least, I hope this book cuts through much of that misinformation and provides some clarity.

One analogy I've heard concerns the old saying 'a chain is only as strong as its weakest link'. Many people think of Grenfell and automatically think of flammable panels. But this is far too simplistic, and the analogy of a weakest link is not just a poor one but hopelessly inadequate. Worse than that, it's factually wrong. Because this isn't a story concerning a chain with one failed link. It's a chain made of failed links. If it was just one weak link, we could re-assure ourselves that 'fixing' this particular aspect would solve all our problems. In reality, multiple links failed and if just one had performed to standard the size of the tragedy would have been curtailed, if not avoided altogether.

If we just put aside the most obvious and visible matter of covering buildings in flammable materials, there are a number of other variables: Adequate fire stopping around the windows may have prevented the fire from penetrating the gap and getting behind the panels. Nor would the fire have so easily re-entered the flats above through that same gap. Efficient fire breaks between the panels would have impeded the vertical and horizontal spread. If the fire doors were sufficiently robust, the fire may not have entered the lobbies so quickly. The same goes for the many faulty door-closing mechanisms. And so it goes on from failing fire lifts to smoke extraction systems. This isn't a case of one thing going wrong: rather one of multiple failures all at the same time.

Yet that only tells part of the story. In truth this chain of events on the night is one small part of another much longer chain: from testing to manufacture, from design to building control, maintenance and inspections to enforcement. The chain of events of that terrible night in June lie in a chain of dumbed-down standards across the board over decades.

Which brings me back to the main reason for writing this book. The most important thing here is the truth. What were the causes? What are the possible solutions? How do we avoid a repeat? I've seen little attempt to adequately or fairly explain all the nuances of the situation. Often this is the nature of today's media. There has been little appetite for television companies to have an hour-long discussion involving experts in fire safety or building control. Even less in looking at the history of how these have evolved over the years. Instead, what we get is a sound-bite from a politician who has little expertise in the subject or, even worse, a celebrity with none. The main danger, however, is people with political agendas attempting to frame the questions in a simplistic and biased way.

How am I qualified? In truth, any expertise I may have is limited and only covers one small part of this incident. I have no knowledge of building control or regulations, both vital components of the tragedy. Nor am I an expert in fire safety. A subject the public and, at times it seems, the inquiry assumes erroneously all firefighters are knowledgeable about. What I have is over thirty years' experience in London. Most of that time was as a firefighter, ending my career at the rank of Watch Manager, or Station Officer, relatively low down in the great scheme of things. I also spent nine years as an instructor, training new recruits, many of whose names I recognised as attending Grenfell while I researched this book. Some I have spoken to. Throughout my career I attended, visited and trained in high-rise blocks more times than I can remember. Ironically, given the events, high-rise procedure is one of the most common scenarios firefighters encounter both in training and in practice.

For my own part I had very little to do with the incident. I attended the next day with my crew and spent about eight hours feeling rather helpless and hopeless, not doing very much at all. The tower was still smouldering with pockets of fire inside the building requiring water throughout the day. Debris littered the area. Some pieces caught my eye. With hindsight these must have been parts of the panels but with only one face of metal sheeting surviving. Attached was what was left of the inner core, expanded like an enormous, blackened foam bubble. When I touched it with my gloved hand it disintegrated into a cloud of black dust which blew away in the breeze. 'How is this even possible?' I thought to myself.

If the broken chain is a poor analogy, can we offer a better one? Let us take the airline industry. Not just one type of plane or one particular carrier. The entire industry. At one end we have a simple bolt. One of many thousands of components that go into a plane. The components get made to a standard, then tested: quality controlled. Fitted together with other components. From the simplest nut to the entire finished plane, multiple tests and inspections. When completed and in operation there are more regular tests, maintenance and inspections. Pilots and crews are trained, and, again all subject to inspections, re-training, re-testing.

Now let us imagine we dumb down the standards of production of our bolt. We will likely get away with it for a while. Later quality control, inspections and maintenance will pick it up. Engineers will 'tut' and bore their younger colleagues with stories of how things used to be done. Even if we dumb down the standards of production of multiple components this will just result in planes being grounded. But what if we also dumb down standards of maintenance? And then inspections. What if we then dumb down the powers of enforcement agencies? What if we spend thirty years or more dumbing down standards throughout the whole system? Then imagine we lose a large number of engineers and those responsible for enforcing safety, either through retirement or underfunding. Whilst all this time those in positions of authority within the industry and politics simply dismiss or deny there's a problem. Despite multiple warnings.

It's easy to guess the likely consequences. Quality would fall, poor materials would be let through by poor testing and inspections. Not just one bolt but a host of different products and materials. Poor maintenance would compound the problem and tick-box or non-existent inspections would let it all through. Near misses would increase, followed by the odd crash. Warning signs will be dismissed or misidentified. Far better to blame a component or find a scapegoat than admit the entire system is at fault. Enforcement agencies, underfunded and now toothless, will struggle to hold back a wall of sub-standard practices. One day a major disaster occurs. When that happens our dumbed-down and toothless enforcement agency will be forced to look at the specific cause of that incident. A bolt will be blamed. Perhaps a specific manufacturer. An inspector at a particular point in the process.

Or maybe the council on whose patch the plane crashes. Alternatively, the government in power at the time. On this point, it might well be that governments of a particular political persuasion were more to blame than others. But, as we shall see, Grenfell involves all governments of all persuasions over the last thirty years.

The purpose of this book is not to deflect blame from the Fire Brigade, although I believe much of the criticism in the press has been ill-informed and unfair. I will point out what I think were mistakes when we come to it. However, the biggest concern for me is that, for many, finding someone to blame will be the end of the matter. This book will argue that this is very far from the truth. Regardless of which individuals or organisations broke laws or guidelines, it will not change all the underlying causes if a few people go to prison or large corporations are fined, although that may very well prove to be warranted. The problem is a deep and wide systemic one from the manufacture of materials, testing of those materials, construction, building control, fire safety all the way through to enforcement.

We have dumbed down multiple links in the chain. Each link is made up of thousands of individuals, companies and organisations. Blaming a handful of such 'stake-holders' will solve nothing. What of our pilot and crew in this scenario? They are trained to land a plane in an emergency. Should the wheels fail to deploy, an engine or two fails or some other malfunction occurs they can, and sometimes do, land the plane safely. It's a bit more difficult if multiple things fail at the same time. Should the rudders jam, the wheels fall off and all the engines stop working all at the same time, then the situation becomes far more difficult.

I will argue in this book that this is a fairer analogy. Multiple things went wrong at the same time at Grenfell: panels, fire stopping, windows and fire doors all failed to behave as they should. At the same time those aspects designed to assist firefighting, the fire lift and smoke extraction system, also failed to work adequately.

Which brings me to what I think is a major point worth noting. The response to our hypothetical aeroplane crash might be to insist planes are fitted with sprinklers and passengers allocated parachutes. Perhaps an emergency plan for those with disabilities. I just want to emphasise that, no matter how sensible these suggestions may or not be (when applied to high rise buildings), it is far more important that we ensure we don't build aeroplanes in such a way that the wings, tail and engines fall off at the same time. It would also be sensible to ground those planes until the fault was rectified. And for that work to be done in a relatively short time frame.

The fact that we have not resolved these questions in residential high-rise buildings in six years is indeed a scandal. But let us at least identify the main underlying problem. Because I can't help feeling that every time the focus is on sprinklers, stay-put policies, a particular political party or any of the myriad of other questions surrounding Grenfell, we are in danger of losing sight of the main problem to be resolved that underpins all others.

It is this: the widespread and deep dumbing down of standards across the board from manufacturing, testing, building control, inspections, maintenance, fire safety legislation, enforcement agencies and emergency services. Some of these may well be more important than others in this case. But to ignore this and focus solely on sprinklers and the stay-put policy, or use it as a political weapon, is to do a great disservice to the public.

This book is dedicated to those firefighters who were thrown into this situation and faced it professionally and with courage. It is also dedicated to the seventy-two people who lost their lives. It is important to note that this is just one firefighter's perception but, hopefully, some at least will find it useful.

Part I

A Timeline and Career

Beginnings

It was difficult to decide where to begin this sorry tale; perhaps with the development of aluminium composite panels, ACM, in the 1960s. However, panels are just one small part of this story. We could begin with the construction of tower blocks in the 1960s. A significant drive produced more homes in the 1960s and 1970s than at any other time. Tower blocks were an important component of this with the construction of Grenfell Tower being completed in 1974. Yet fires in such blocks were dealt with every day for years without tragedies such as Lakanal in 2009 or Grenfell in 2017. How did we get through those decades without compartmentation failing so catastrophically?

As this book is not aimed at experts, I will explain terms as I go forward. Compartmentation is the sub-division of a building into smaller units. Those units are protected from the spread of smoke and fire. One important example of how a flat, or escape route, might be protected is the fire door. Thus a fire (and smoke) should be contained in a flat for a specific time, allowing people to escape, limiting damage from smoke or fire spread, and keeping people safe in other parts of the building.

It might be more relevant to Grenfell to begin in 1984 with the part-privatisation of enforcement. The Building Act (1984) introduced Approved Inspectors, independently monitored and regulated by the Construction Industry Council Approved Inspectors Register (CICAIR) to carry out building control work in England and Wales. The existing 350 pages of building regulations were replaced by 24 headline standards. Rules and guidance concerning fire were contained in a document known as *Approved Document B*.[1]

It's important to note at this point that firefighters are not building control officers. A fact I felt that some in the media never quite grasped. There is no reason why a firefighter would know anything at all about Approved

Document B. Officers in fire safety would be aware, but firefighters are not fire safety officers, another fact that appeared lost to the media and, even at times, to the inquiry. I had heard of the document but only because I'd been seconded to Fire Safety for a couple of months later in my career. Approved Documents provide guidance for meeting building regulations with Part B relating to fire safety, including means of escape, fire spread, structural fire protection and fire service access.

Whatever the concerns there may have been about deregulation or commercialisation of building control, there were no Lakanal House or Grenfell Tower type fires when I joined the fire brigade in 1986. As noted in the introduction, one of the main arguments in this book is a systemic dumbing down over several decades. The other is: how did we get to the point where firefighters felt so shocked and apparently ill-prepared at Grenfell?

For this reason, I have decided to list the relevant questions chronologically alongside my own career. Many of these I only became aware of after Grenfell. The best source of information for those interested is the Grenfell Inquiry website. In particular, I have found three books extremely useful and educational: Tony Prosser and Mark Taylor's excellent study, *The Grenfell Tower Fire: benign neglect and the road to an avoidable tragedy*. The sub-title, *benign neglect and the road to an avoidable tragedy*, is a description I can agree with fully. For those looking for a more detailed exposé of the litany of mad, bad and dangerous decision making in industry and government, then you can do no better than Peter Apps' best-selling book, *Show me the bodies, how we let Grenfell happen*. Lastly, Gill Kernick's *Catastrophe and Systemic Change, Learning from the Grenfell Tower Fire and Other Disasters*.

Training school

I can't remember when exactly I decided I wanted to be a firefighter, but it would have been around the time I was doing my O-levels (one of the last to do so before GCSEs became ubiquitous, I believe) at the age of 16. So convinced was I that I entirely gave up on thoughts of university and any other career. Consequently, I left school at 18 with just one aim in mind. I think what appealed was the idea that I might be doing something I considered worthwhile. I certainly had no inclination to work in an office.

The wait even to apply was as long as it is today, and I found myself doing a range of temporary jobs while I waited for an interview. When the chance came, there were apparently 15,000 applicants for 300 places, a ratio that has been fairly common with each recruitment drive since. Months dragged by and I spent the best part of a year trying to get into shape. At the time there was a minimum chest expansion and restrictions on height. The latter requirement was discarded long ago.

This meant lots of running to ensure a chest expansion of two inches (another requirement long gone). Miles and miles I pounded all through the year. Another requirement was to be able to perform a 'firefighter's lift', put someone on your back and carry them 100 yards in less than a minute. Every chance I got I coerced my friends to help me practise. Odd now to think firefighters have not been allowed to practise this for many years. In fact, I was told in no uncertain terms that I was not to teach this at all when I was a trainer much later.

After several months of delays, I grew slightly despondent and began to wonder if I would ever get in. I started a small business as a DJ and applied to what was then called the Enterprise Allowance Scheme. Finally, the process moved on and, a year after leaving school, I found myself in an interview, having passed the medical. I breathed a sigh of relief at the chest expansion stage and was slightly put out when the assessor waved us through the fire-fighter's lift assessment after just a few yards. After practising so hard for so long I would have liked to finish the 100-yard course.

Then more waiting. I was just beginning to wonder which way my life would turn when, in the late summer of 1986, I received two letters on the same day. The first accepted me on the Enterprise Allowance Scheme and a career as a DJ beckoned. The second letter was from the London Fire Brigade. I remember standing there with an acceptance letter in each hand and wondering for a second. Not that there was any doubt. A dubious, unstable career in the entertainment industry went in the bin.

My official start date was 22 September 1986 but a training course was not available for a few weeks and so a month and a half was spent making tea and filing at headquarters. The course, when it started, was twenty weeks long. We shall see how this has been cut down to just eleven weeks today. I recall lots of marching and discipline. There is no longer a requirement to march but it is a perceived reduction in discipline standards that many older hands notice. Like many recruits before me, I remember vividly the first time walking under the famous arch at Southwark Training Centre, a short walk from London Bridge.

One notable feature of the course that was later dropped (certainly by the time I returned as a trainer) was that there was at least some fire safety training. It may have only been a lecture or two, but it was better than nothing. On the other hand, other topics were to greatly improve. Road traffic collision training was confined to cutting up an old metal locker with a pneumatic hammer chisel. A far cry from the specialist cutting tools and several old cars we would get to cut up for each course years later.

I passed out of training in the spring of 1987 and was posted to Green Watch, East Greenwich in South-East London. It never occurred to me that I would return to Southwark to train recruits. One difference I observed between those two periods was that the type of trainee had changed. In 1986 a significant proportion of recruits were ex-military, tradesmen or young but fairly sporty people like me. By the 2000s it was not uncommon to have an entire course consisting of people with no practical experience of working with tools or other physical activity.

I mention this merely as an observation. It is not necessarily a bad thing. There is no reason why someone with no experience cannot learn how to use certain tools. But I can't help noticing that it is a very different experience training a squad full of ex-service and tradespeople compared to a squad who have never picked up a hammer or used any basic tools. I would suggest that, in an effort to diversify, the Brigade forgot the essence of a major part of the job, a practical, physical job that requires a level of dexterity, physicality, fitness and strength.

A fire station consists of four watches, or shifts, Red, White, Blue and Green. At the time a two-appliance station, as East Greenwich was back then, had a Station Officer in charge with a Sub Officer and Leading Firefighter assisting and ten firefighters. By the time of Grenfell, the watch strength had fallen by one and the ranks changed to Watch Manager and Crew Manager. Green Watch East Greenwich was full of what some might call *characters*. All were very experienced, and I was the youngest by nearly twenty years.

What the public know as fire engines we call 'appliances' or 'pumps'. A fire station is classed as one- or two-appliance, depending on how many fire engines are based there. Some of these stations also have specialist appliances. Various aerial appliances include turntable ladders, hydraulic platforms or aerial ladder platforms. Command units assist at large incidents such as Grenfell. One specialist appliance worth noting is the Fire Rescue Unit as that carries, among other things, extended duration breathing apparatus, vital if firefighters are to penetrate fires: the normal standard duration sets carried on pumps have insufficient air for this.

In those first years of my career, it was common practice for crews at a high-rise residential block to go straight to the flat concerned to investigate. It was drummed into us never to go empty-handed and so you'd often see firefighters with an extinguisher, a sledgehammer or a long line. The latter was a 30-metre length of line carried in a bag for hauling equipment aloft (a line being a length of rope cut to a specific length). The sledgehammer was far more useful in breaking in a front door than the ubiquitous firefighters' large axe seen in films. On more than one occasion the initial crew attending would make entry and extinguish the fire very quickly with an extinguisher and render further assistance unnecessary.

One important concept, the 'snatch rescue', was then standard practice. This is when a firefighter, without breathing apparatus, attempted a quick rescue in a situation where he or she deemed the potential of a successful outcome outweighed the risk. A common scenario might be a firefighter making access for a breathing-apparatus crew and seeing or hearing a casualty within easy reach. Instead of waiting for the breathing-apparatus crew to don their sets, you might judge it possible to crawl in safely and pull the casualty out with minimal risk.

As the years rolled by such concepts became frowned upon. To be fair the Brigade was trying to address a culture that accepted a level of smoke inhalation as part of the job. More than once in my early years I was encouraged to enter thick black smoke without breathing apparatus in situations where it was completely unnecessary, such as a car alight in an underground car park.

Time and again after an incident I'd be coughing and sneezing black mucus for hours after. No wonder the life expectancy of retired firefighters was so low when I first joined. Slowly the message that wearing a breathing-apparatus set was for our benefit got through. This cultural change is to be applauded. By the end of my career, firefighters regularly dealt with car fires in open air using breathing apparatus, something one never saw when I first joined. Such was the culture over thirty years ago and many have suffered healthwise because of it. It is undoubtedly a positive development that that culture has changed. The danger is that grey areas become firm lines that slowly get redrawn and moved further away from what some would call common sense and a reasonable risk.

6 The Grenfell Tower Fire

A level of 'operational discretion' is sometimes necessary which we saw countless times at Grenfell. There we saw many examples of firefighters going against procedures to perform rescues. I am in no doubt that, in other circumstances, some of these firefighters would have been disciplined for the very necessary actions they took that night. The high media profile ensured that no such action took place. This is a theme we will return to: grey areas and decisions being influenced by concerns other than what constitutes a reasonable balance of risk versus reward.

I attended my first fatal incident within three months. Two young lads, around my age, driving fast on the Blackwall Tunnel Southern Approach. They tried to beat two lorries to the bend, racing between them in the middle lane. They made it, but sadly not the bend just a few yards ahead. Hitting the central reservation, they flipped over which must have split the fuel tank. I recall seeing the pall of thick black smoke rising from the flyover a few hundred yards away from the fire station's forecourt. Seconds later the call came in. On arrival the car was ablaze.

For those entering similar jobs and who have never experienced such things, it's always a concern how you might respond. We were confronted with a horrific sight. The driver was clearly already dead, and I will spare the reader the details about why that was obvious. A worse sight was the passenger who had been thrown clear. Or perhaps he had crawled out.

A lorry driver was in tears, having tried to put the fire out with the only thing he had, a dry-powder extinguisher. I had to try to coax the injured man to sit down. A very difficult thing to do when there is no part of the body that has not been affected by the fire. Again, I will spare the reader. Suffice to say that in my 31 years that remained the worst thing I ever saw. Somehow the man managed to survive a day or so before succumbing to his injuries.

I recall a man taking two young boys over a walkway to look down on the incident and incredibly taking a camera out of his pocket (this being long before mobile phones). We put sheets up to block his view. Drivers on the other side started rubbernecking and nearly caused a crash. A coach driver was dragged across the central reservation by an officer. Someone said he'd failed a breathalyser. The officer grabbed him by the scruff of the neck and forced him to look into the car. The man went white as a sheet.

No doubt the policeman would get the sack now, but I can't help thinking that had more effect on the coach driver than any number of points on his licence or fine. It certainly had an effect on my driving. I remember being relieved that I had got through that incident and coped okay. I knew someone who had left the police after a just a couple of years because of a particularly grisly road traffic collision. In truth, I was too busy to think about it at the time, either dealing with the casualty or moving equipment about. I considered how precarious life could be. One wrong decision made by you or another driver could change everything.

I did have a strange experience that I have never told anyone about. For a few weeks every now and again the smell would return. I didn't feel upset about the incident, there were no bad dreams and I didn't think it had affected me in any way. Yet a handful of times, randomly, I would detect a distinctive smell, the same smell from that incident. I can't explain it and it stopped after a few months. An odd thing, and I only include it here to illustrate that the mind is a curious thing and we often can't explain or always control it. I wonder how many survivors and firefighters are plagued by such memories from Grenfell.

In terms of dangers to firefighters it is important to note the subject of cladding never came up. The closest was possibly sandwich panels, large sheets of metal with a core of often flammable, toxic insulation and often used in the walls of large out of town commercial units. If the core becomes exposed, rapid fire spread can ensue, giving off significant amounts of toxic smoke.

One example that I do recall being drummed into us was the dangers of building collapse. We learnt the basics of the signs and symptoms of a collapsed building in training, and this was re-emphasised at station in a lecture. Dropped arches, cracks in walls and beams pushing out brickwork to name but a few.

Another risk was cylinders. In a fire the liquified gas can expand causing the cylinder to fail. Whilst they should have a pressure relief valve, if this fails, or the temperature and pressure increase too quickly, the result could be an explosion, sending shrapnel from the cylinder over a large area. I recall a fire in my first year involving a large outbuilding full of scores of cylinders and hiding behind a wall trying to direct some water onto it as every few seconds a cylinder exploded. The nearest I ever came to what I imagine a war zone might be like. It was interesting to see the relief valves of those cylinders outside the building actuate and the escaping gas catch light and shoot a flame several feet into the air. Then another large explosion as a cylinder failed had us cowering behind the wall again.

The main risks to firefighters, so I was always told, was a flashover or a back draught. A back draught is when fire has been bottled up in a compartment for some time. Fire requires fuel, oxygen and heat (often called the triangle of fire). If you imagine a fire that has consumed nearly all the oxygen but still has sufficient heat and fuel, wait long enough and that fire will burn itself out and the room cool down. However, should you open the door or break a window at that specific moment, you allow air to enter and mix with the hot, unburnt gases. If there is still a flame present, or the gases are above their auto-ignition temperature, the subsequent ignition of those gases will exit the opening with potentially explosive force.

Needless to say, you don't want to be standing in the doorway when that happens. Thus, firefighters use a set 'door procedure' to enter a room. The first breathing-apparatus crew to enter the kitchen at Grenfell used this very procedure, crouching down with one firefighter controlling the door handle, using either the door or wall as protection depending on whether the door opens towards or away from you. The second firefighter controls the hose, taking a quick look at conditions and, if warranted, sending two quick bursts into the gases at ceiling level. If the decision is made to enter, you get in quickly, away from the opening and keep low.

Signs of a back draught might include yellow or brown smoke. Generally, the darker the smoke the more incomplete the combustion, the more unburnt gases are present. A type of 'breathing effect' might occur around gaps in doors and windows, with the smoke puffing out and being pulled back in. Also blackened windows and no visible flames.

To understand a flashover, it is best to imagine a growing fire producing a large amount of radiated heat. As the temperature increases, everything in the room starts to pyrolyse, decomposing and giving off gases. One definition is: the near-simultaneous ignition of most of the directly exposed combustible material in an enclosed area or, in other words, a fire in a room suddenly becomes a room on fire. One sign might be a rapid increase in temperature and items in that room giving off smoke. Another might be tongues of flames in the rolling gases at ceiling level as they reach their auto-ignition temperature, this being the lowest temperature at which a substance ignites without an ignition source. Some materials have a very low auto-ignition temperature which is why it's generally a bad idea to cover buildings in such materials.

Thankfully in those first years I never encountered either of those. There was one very famous fire at this time that did involve a flashover. The King's Cross fire occurred in my first year, November 1987. A fire began beneath an escalator and soon the entire escalator was alight, sending superheated gases up into the ceiling of the shaft. The ensuing flashover sent a jet of flames up the shaft and into the ticket hall. Thirty-one people perished, including Station Officer Colin Townsley. A public inquiry resulted in smoking being banned on the Underground. A welcome move, but there was the nagging thought of why we needed a tragedy to implement sensible policies.

It was a painstaking process for fire investigators and forensic scientists to piece together the cause of the fire and identify the casualties. One victim was not identified until January 2004, nearly seventeen years later. Such is the professionalism and perseverance of those whose job it is to identify people. That's something to bear in mind when conspiracy theories swill around about Grenfell, claiming either that hundreds perished and there was a huge cover up or that the building was empty and the whole thing was a 'false flag'.

It is difficult to recall when I was told about the stay-put policy in residential high-rise blocks. It's one of those things I feel I've always known, yet can't remember ever being taught about. It was a policy that belonged to the building. We attended high-rise premises and they had a stay-put policy and that was that. The only reason I ever remember being told from my earliest years was that compartmentation meant the people did not need to evacuate. It was only much later in my career that anyone ever expanded on the reasons. I certainly never had a reason to question it.

Compartmentation did indeed seem to work. In my early career I never attended or heard of a case where it failed so catastrophically as it did at Lakanal and Grenfell. As compartmentation worked, so did stay-put. Incident after incident I attended at high-rise blocks people remained in their flats. I don't recall ever attending a block where stay-put was not the building's policy. Yet, at the same time, the advice to people in other types of domestic buildings was to get out and stay out.

Get out and stay out vs stay-put

Most of us remember being told at one point that if there's a fire 'get out and stay out'. Don't go back for your money, possessions or a beloved pet. This is indeed still the mantra. Generally, this applies to all domestic dwellings: houses, bungalows and flats in converted premises. A second type of building includes commercial premises, places of work and those of entertainment: factories, shopping centres, schools, hospitals, football stadiums and many others. In these premises there is generally an alarm system and, importantly, people employed to assist evacuation. Why then are residential high-rise premises different?

Let us look at the first two types of premises in a little more detail. A person in a domestic house will, if able, self-evacuate before the fire brigade arrives. If they cannot achieve this then it becomes a rescue. Firefighters in breathing apparatus will enter the premise and search and rescue. This distinction between a rescue and an evacuation is important, despite the words often being used interchangeably. It does not help that, at times, even firefighters conflate these terms.

At a commercial premise the owner has a responsibility to train staff in what to do in the event of a fire. This is underpinned by specific legislation. Staff have a responsibility to follow those guidelines. There are regular drills in which staff are trained to respond and other staff are designated to assist and direct any evacuation. In schools, teachers will guide the children to a playground or field away from the fire. In a nightclub, designated staff will direct clubbers out of the building. Football stadium stewards will evacuate the supporters from the stadium. In a factory or office block staff will be trained to evacuate to a safe 'fire point' and fire marshals or wardens will direct the operation. An employee who does not comply can be disciplined, re-trained and eventually sacked. In every single one of these cases the evacuation is carried out by on-site, trained staff before the fire service arrives.

It cannot be emphasised enough just how important this last point is and it's worth repeating. Evacuation is carried out *before* the arrival of the Brigade. We will come back to that in a moment. When an area near to a large fire requires evacuation then the officer-in-charge will often request the attendance of the police to carry this out. Firefighters may assist in this but have no powers to move people on and rarely have spare personnel to get involved, certainly not on any large scale.

The third category is the residential high-rise. This is different from both a domestic house and a commercial premise. Unlike a commercial premise, there is nobody on-site to carry out an evacuation before the fire brigade arrives. There is no one to manage and direct people. The occupants of a high-rise will not be subject to any discipline by an employer should they not comply.

There is nothing stopping the building owner from providing on-site persons to co-ordinate an evacuation before the Brigade arrives. A number of concierges would be required, working a shift pattern, and their wages would have to be divided between the flats. Let's imagine four persons on £25,000 a year. A large block could spread the cost across more flats. One hundred flats might contribute 'as little' as £1,000 a year or nearly £100 a month. A small block of thirty flats might have to pay several hundred pounds a month extra.

Another alternative would be to install an alarm system covering the whole block instead of single alarms in individual flats. There are two main reasons why the vast majority of blocks (well over 99 per cent in my experience) do not have an integrated or communal alarm system. Firstly, experience has taught us that after a while people often ignore alarms, especially when they occur frequently. Having attended fire alarms in nursing and student accommodation it becomes apparent that after several false alarms a percentage of students simply ignore them.

The second reason is the stay-put policy. In theory, compartmentation makes an evacuation unnecessary. It was later in my career that I heard the following explanation. Encouraging scores of people to open their doors at the same time just as a fire is escalating might have consequences, especially if this is at the same time that the Fire Brigade is attempting to gain access to the fire compartment. Furthermore, placing hundreds of people on the, often, only staircase just as firefighters are coming up and making entry into a fire compartment might not be the best option.

Thus the vast majority of residential high-rise buildings in the United Kingdom operate a stay-put policy. This policy worked for the decades before and much of my career. Whilst we struggle with why it failed so disastrously at Grenfell, and at Lakanal House a few years before, it might be worth considering why it worked for so long? And what changed?

If some readers started this book convinced that stay-put is always a terrible idea I must ask you to keep an open mind. Not least because prisons, ships, hospitals and care homes all have a de facto 'stay-put' policy. Leaving those examples to one side for a moment, let us focus on residential high-rise buildings. It is also worth noting at this point that the 'stay-put' guidance does not simply say 'stay put'. What it says is 'Stay-put *unless* you are affected by fire, heat or smoke in which case get out *if* you can'. If you cannot, this then becomes a rescue scenario, a nuance the press and other media often failed to make clear.

What is a residential high-rise?

One definition, from a US former Deputy Chief Officer defined a high-rise as a building taller than your tallest aerial appliance.[2] Another definition defined it as a building 'that has floors that are inaccessible by external means'. In the UK it is simply a building that has floors more than 18 metres above ground level. This usually means buildings of over six floors, 6,500 of which were built between 1945 and 1990.

Commercial high-rise blocks will have an alarm system and staff on site to assist evacuation. One building I attended at Canary Wharf had a 'phased-evacuation' system. This is where they will evacuate certain floors first. This might mean the floor on which the alarm sounded and the ones directly above and below first. A residential block would usually have no such alarm system. This is deliberate, a point many a politician seemed unable to grasp after Grenfell, with some questioning why there was no fire alarm.

Residential high-rise blocks are recommended not to have a communal alarm system precisely because of the stay-put policy. Instead, flat owners are recommended to have individual smoke alarms because you are only supposed to evacuate the flat affected. The whole point is to avoid the situation where everyone is opening their doors at the same time as this might allow smoke and heat spread from the floor the fire is on. Also, it is to avoid putting everyone on the only staircase just as the fire brigade is arriving and trying to gain access to and fight the fire.

The only means of access and egress is via that staircase. Most blocks in London only have one such staircase. In an ideal world it might be preferable to build blocks with two staircases. However, it is worth pointing out that blocks with one staircase were perfectly safe for many decades in terms of fire-fighting.

Two features of residential high-rise buildings are worth noting. Firstly, fire lifts. There are four types: Firemen's lift; Fire-fighting lift; Fire-fighters' lift; and modernised lift for fire service use. There are some technical differences between them, but they allow a firefighter to call the lift to take control of the lift via a fire-control switch, usually operated via a panel above the lift doors on the ground floor. The lift returns to the ground floor and allows fire-fighters to control the lift from inside the car. It will no longer respond to those on other floors pressing the button in the lobby. The lift can now be used to carry fire-fighters and equipment from the ground floor to the bridgehead.

Secondly fire doors prevent the spread of fire for a specified time, usually either 30 minutes or 60 minutes, depending on the fire door rating. They have a number of features: a self-closing device; intumescent strips and cold smoke seals to top and sides; substantial hinges, usually three sets; a gap of 2-4mm between door and frame; and fire door signs clearly displayed. They not only protect individual flats but also the escape routes, most notably the lobbies and staircase.

The more fluid and practical definition of the US former Deputy Chief raises the subject of aerial ladders. However, it must be remembered that the nature of London makes some buildings and streets inaccessible to many of the very large aerials. In addition, it is far more difficult to put fires out or to rescue people from a ladder outside the building. Far easier and safer to fight fires from inside. Lastly, there comes a point when the laws of physics simply won't allow a ladder big enough to reach the top floor of a very tall building. Without getting into exact measurements there comes a point when the building relies on something other than ladders to rescue people.

All this relies on compartmentation. We recall that this meant dividing a building into small units, each protected from smoke and fire spread from another part of the building. We will now turn to that in more detail.

Compartmentation

Creating a fire-resisting compartment means that the doors, walls and floors of each flat resist the passage of fire and smoke and maintain their structural integrity for a specific length of time, usually between 30 to 120 minutes. In practice, this means that if you are in one flat with a fire next door then, at a minimum, you have two, thirty-minute front doors between you and the fire. The walls and floors will generally be much higher rating. This gives enough time for the Fire Brigade to arrive and deal with the fire. Should the fire break out of the original compartment this would normally occur in one of two ways. Firstly, via a window being open or eventually failing and smoke and flames spreading up to the flat above. Secondly, via the front door being left open or failing and the fire spreading to the corridor.

In these scenarios the crews would attempt to evacuate that floor or the one above. Although whether these are evacuations or rescues is open to debate. It probably does not help that fire-fighters would often use the terms synonymously. As we shall see, the distinction between the two is actually quite important and I would argue that they are properly defined as rescues.

We can see here that the doors are an essential component of this system. Should the front door of a property fail or be replaced by a door that is not a fire-door then the lobby is at once compromised. As this is usually the only route to the staircase for the other flats then one can immediately see the danger. That is bad enough but what if those other flats have also replaced their fire-doors? In that case our one-hour protection is no longer present which compromises the safety of the entire building.

Equally important are the doors to the staircase. Not only are these the main access for fire-fighters but, often, the only means of egress for residents. It is thus absolutely paramount that the staircase remains safe. Hopefully, the reader will see immediately that the safety of the entire block relies on high standards of building, building control, maintenance and inspections. It also requires adequate enforcement. All are things that this country was once considered good at.

One officer, retiring a few years before Grenfell, with over 30 years' service, the majority as a Fire Safety Officer, was concerned enough about what he described as the 'modern Risk assessed approach' that he came out of Fire Safety altogether. One of his main concerns followed the increase in 'right-to-buy'. This wasn't a political view but rather driven by people's tendency to change their front door. This is no problem in a single domestic house. But in a block of flats this act immediately 'compromised the Fire Protection of the buildings by removing the one-hour fire doors (FR) that had been in place for years'.

> I am not certain that Landlords had the powers or knowledge to stop the UPVC door issue and insist on fire resisting doors. Flat owners were largely ignorant of it. Later local authority building control lost their exclusive role and Approved Inspectors were introduced. Some were good, others were dire. The design of most blocks of flats was of a central core of lifts stairs and risers (not always central in the building)

with hallways off this to the flats. The stairways were protected from the hallway by a fire resisting door. The hallway was generally the firefighting lobby.

Doors to the flats were fire rated so crews would set into the riser which was in the hallway and start up their breathing-apparatus sets. They could then enter the flat of origin and fight the fire. Once these firefighting lobbies were compromised by non-fire rated doors the crews had no option but to start up elsewhere. The most obvious place is at the next nearest dry riser outlet, the floor below. This became operational policy after a few near misses and accidents.

What should have happened at that point was Brigades should have asked why lobbies were not working. There was also no legislation for the fire services to regulate on residential property. These have always been housing act issues. Fire Precautions Act 71 and its 2 designating orders only covered Hotels and offices, shops and railway premises. We used to do a goodwill inspection of dry rising mains each year but these were stopped in the mid-80s.

Another problem he saw was the increase in UPVC double glazing. Whilst this improved matters around condensation in a fire, the frames would fail quicker than those made from metal or timber. He continues: 'It became practice to set into dry risers on the floor below the fire, this meant that hose had to run up the staircase, this compromised two floors of fire lobbies and the staircase, as well as obstructing the only means of escape. Later cladding was added to the outside of buildings for insulation and modernisation purposes, with the sad results.' Unfortunately, some seemed unaware then, and now, of the importance of fire doors, windows and external surfaces in residential blocks. As well as the need to keep the staircase free.

A further problem the officer above noted was the change from statutory national exams and reduction of commitment to the Fire Service College: 'Firefighters, and in particular their leaders, the Watch Officers and middle managers, lost the background knowledge of basic building construction, fire safety and fire protection. I have accompanied crews on familiarisation visits to modern constructions and asked questions about fixed protection systems to a sea of blank faces, including the officers.' Importantly, he stated, 'My first station was in a busy North London Station, we had well over a hundred council blocks on our ground as well as numerous private ones, high-rise fire-fighting was bread and butter. I must have attended Grenfell numerous times, although I don't specifically remember doing so. I am of the opinion that stay put was not a policy, it was a *fact* that when built there was no need to even consider evacuation. No fire ever left the flat of origin and at worst the neighbouring flats were only affected by a little smoke.'

What then can we glean from experienced fire safety officers like this one? Firstly stay-put worked as well as the buildings did for several decades. But small changes have consequences and the reasons for certain practices are sometimes forgotten. No one seemed to care enough to monitor fire-rated doors which were vital for the fire safety of a block of flats. Given how fundamental a front door is to compartmentation, it is surprising little heed was taken later when maintaining, refurbishing or inspecting these same blocks. The Brigade's response was to change their policies rather than fight against the increasing dumbing down of standards. Perhaps this was in part due to underfunding and undermining their role as an enforcement agency. Whatever the case, it shows how we as a society allowed important lessons to be forgotten or ignored.

New station

After a couple of years, I moved to an inner London station where, I'm afraid to say, I experienced the worst part of my career. Recently, in 2022, an independent review delivered a damning verdict on the Brigade culture. Nazir Afzal, the former chief prosecutor who conducted the review, labelled the brigade 'institutionally misogynist and racist'. I personally found this rather ironic and not because I had any doubts over the findings. Whilst I don't completely recognise the picture it painted I am not as naive to think that, just because I did not witness this level of inappropriate behaviour in the last years of my career, it did not still go on.

The reason why I found it ironic is because in my experience the culture had massively improved over the thirty years I spent in the job. Back then certain watches had a reputation. I was unfortunate enough to find myself on one such watch. So bad was the situation that, if I had not been recently married and become a father for the first time, it is quite likely I would have left the job I had been so eager to join. The environment was hardly conducive to an effective fire-fighting team. Quite the opposite.

Suffice to say the level of general bullying and inappropriate behaviour was every bit as extreme as anything Mr Afzal found in 2022 and, in many cases, a whole lot worse. In my youth and inexperience, I thought I had to just suck it up. Two years later, I got a good lesson on how I should have handled it. A firefighter I knew from East Greenwich was posted to the watch. Three days later he was gone. He had witnessed the behaviour and gone straight to the Station Commander, demanding to be transferred or he would put it all in writing. Very telling that back then there were officers who would rather sweep such behaviour under the carpet than deal with it.

Off he went and the watch continued down its dysfunctional, toxic way. It was only the arrival of a new Station Officer who turned it around. I got quite a lot of good tips from watching how he handled the situation. This book is not the place to cover such things, but I raise it to demonstrate one important thing. The culture I joined had been completely transformed by the time I left. In so many ways this has been for the better. Watches like the one I experienced were relatively rare even then. At the end of my career that type of behaviour was confined to a small number of individuals and was less extreme than what I had experienced when I joined. This does not detract from any of the findings in the recent report. It just needs to be acknowledged how different things are from when I first joined.

A first warning sign?

In 1989 a pilot cladding system was installed at Knowsley Heights, an eleven-storey block on Merseyside. Two years later a fire occurred which spread to the cladding system. The subsequent report raised the question of using flammable cladding and called into question the type of tests used on such materials. However, the focus rested on inadequate fire barriers. Evidence shows that the result of the investigation was to 'play down the issue of the fire'.[3] Later an expert witness to the Grenfell Inquiry stated that, if this had been acted on, the 'entire crisis may never have happened' and 'it is impossible to overstate the importance of what we missed here'.[4]

I haven't found any historical UK cladding fires prior to Knowsley Heights in 1991. I think it's fair to say that fire-fighters at station level in London were blissfully unaware. Flat fires in residential high and low-rise are fire-fighters' bread-and-butter jobs. As the years rolled by, my colleagues and

I attended them regularly. Compartmentation held and 'stay-put' worked year after year.

The reader may be surprised to learn that this incident only became known to me in researching this book. If someone had said 'cladding' to me the first thing to come to mind was sandwich panels. Perhaps this is the 'ground zero' event?, the first notch on the temperature for a frog in boiling water, the frog representing standards of building regulations and fire safety. The slow drip drip of a tap as common sense and principle circle the drain.

Learning curves

One of the things about an inner London station is that it does tend to be busier and so it proved. I briefly attended the Clapham rail crash in December 1988 where thirty-five people lost their lives and 484 were injured. A crowded passenger train slammed into the back of a stationary train outside Clapham Junction and then struck an empty train travelling in the opposite direction. It was my first sight of a major incident. We arrived in the afternoon, many hours after the crash at 8:10am. Crews had ripped up fences and placed them down the steep slope of the embankment to enable people and equipment to get to and from the crash site.

We were ordered to move some equipment and were soon told we weren't needed and sent home. The last thing I remember was a crew struggling up the bank with a stretcher. Someone at the time said it was the last person out. The last casualty was taken to hospital at 13:04 and the last body was removed at 15:45, the latter time being the one I believe I saw. Another inquiry produced more recommendations. Another gate shut after a horse bolted.

Another was the *Marchioness* disaster in August 1989. At 1:46am the dredger *Bowbelle* hit the much smaller pleasure boat *Marchioness*. It sank, taking with it fifty-one people. I have a vivid memory of this for two reasons. The first was the initial confusion. Attending from Southwark to the south side of the river, we drove back and forth and saw nothing in the darkness. We even started to think it was a false alarm although the fact that the call had come from the police suggested it was genuine. It was only going farther along the river that we noticed that the crews attending north of the river had located the spot where the *Marchioness* sank.

Which brings me to the second reason which has some relevance to Grenfell. We were there all night but there was little we could do except look into the water. I was detailed with another firefighter to stand on one of the bridges with a searchlight and a long line, supposedly to throw it to someone if they emerged. A hopeless task. The fire boat went up and down the river, eventually finding one victim some miles away. Those who survived were pulled out in the first few frantic minutes, mainly by the police.

There was very little the firefighters could do, except assist the police in patrolling the bridges and embankment. But that is not how the press portrayed it. In short, I did not recognise the dramatised version of events that appeared in the papers – my first lesson in not to believe what you read in the papers. A small taster for the future with the enormous wall of nonsense that followed Grenfell.

A number of incidents spring to mind from my early years. A man out with friends on a Friday night messing about on a tube platform slipped and unfortunately fell on to the tracks just as a train came into the station. One of us, Steve Rogers if I recall, had to volunteer to go under the train to help retrieve the body. How heavy he seemed once we got him on the platform and half a dozen of us carried him to an awaiting stretcher. I remember thinking how apt the phrase 'dead weight' is and how precarious life could be.

Another incident just inside a neighbouring station's ground. Being on their patch they would normally get there first and provide the breathing-apparatus crew. This time we arrived just before them and were already off the appliance when we saw them pull into the road a couple of hundred yards away. It was difficult to tell by looking through the windows if there was a fire in the darkness. But a quick lift of the letter box resulted in black smoke pouring out.

The decision was made to kick the door in. This was an old-style glass-panelled front door. The breathing-apparatus crew still had not pulled up, so two of us went crawling on our bellies to check the front room. Heavy smoke filled the room with only a few inches at floor level allowing any visibility. Looking across the room I saw two feet, just about made out the bottom half of someone slumped in a chair and shouted 'casualty!'

We managed to pick her up and carry her to the front door. As we got there, hands appeared to take her off us. I held the legs, and feeling my grip loosen and seeing the broken glass by the front door I yelled, 'Mind the glass!' 'Righto,' came the reply but, as they took over, her legs fell to the floor right where the glass was strewn. Again, 'mind the glass!' and once again a reply indicating they had heard the warning. But they hadn't. What they thought they heard was 'put her on the grass', meaning the front lawn. A close-run thing as she narrowly avoided further injury.

Such misunderstandings and miscommunication are common in these situations. Usually, we repeat orders back to each other, but mistakes do happen. If emergency workers make such mistakes how more likely is a member of the public to mishear or misunderstand in an unfamiliar, stressful situation?

I can date this incident roughly as it was during an ambulance strike (September 1989- February 1990). I recall attempting CPR and then she was placed on a short section of ladder as a makeshift stretcher, then in the back of a police van. With no ambulances, the last we saw was the van racing to the hospital with the ladder sticking out of the back of the open doors, a police officer inside holding on for grim life. Sadly, the lady died, and I attended the coroner's where I learned that she had succumbed to smoke inhalation.

The evidence suggested that she had stood up and likely taken a couple of breaths of the thick, black toxic smoke. That was enough and she had collapsed on the chair I found her in. Something to bear in mind when expecting people to evacuate through lobbies and a stairwell full of smoke.

We had attended the same address a couple of weeks before where the lady, clearly confused and unwell, had threatened us with scissors. One major improvement over the years is that any case where someone appears vulnerable would now be referred via a safeguarding report to a senior officer. In this way a more joined-up approach would pick up if someone had come into contact with various emergency services and social services on multiple occasions. Nice to be able acknowledge something that has improved.

Basement fires are particularly hazardous as there is often nowhere for the heat to go except the very entrance we need to go down to gain access. One of the only times I felt fearful was on such a job. It was like a blast furnace, and you could feel your skin tingling from the heat before you stepped up to the open door, even through the thick fire-gear. Yet down we went through the heat. Far hotter than what I experienced in real fire training. Surprisingly once we reached the bottom the conditions were relatively cool.

Another occasion brought home just how quickly the heat levels can change. I had crawled in, again as number two in a breathing-apparatus crew, and at one point attempted to adjust my position as I was dragging hose. I tentatively tried to get on my hands and knees to see if I could get into a crouch. No sooner had I raised my back a few inches than it was like someone had thumped me hard on the back. A literal hard smack. Down I went. I tried again and found I couldn't. We put the fire out and got out quickly. Once outside I realised my neck and ears were red (this was before we had protective fire-hoods). I wore an older-style helmet where the plastic visor lifted over the top of the helmet. It had warped from the heat, twisted out of shape so badly that it had to be replaced.

I also went to many high-rise incidents: shut in lifts, rubbish chutes and, of course, flat fires. Each time stay-put was the policy of the block. I don't recall in those years ever attending a block that did not have a stay-put policy. Nor did we ever need to evacuate a block. In my entire career I think I have seen one block that did not have a stay-put policy and had a twenty-four-hour concierge whose job it was to manage evacuation before the Brigade arrived.

We also carried out inspections. These were not formal fire safety inspections. How could they be? We had no fire safety training. Instead, they were familiarisation visits. We checked the nearest hydrants, the dry riser, tested the fire lift and checked for any obvious defects, such as defective fire doors and fire loading in the communal areas (basically lobbies and stairs should be free of combustible materials and allow a clear path of exit). The vast majority had one staircase.

I had an interesting job across the river in Shoreditch or Whitechapel. A shop with flats above. The fire had ripped through the back storerooms and into the upper floors. I found myself leading a breathing-apparatus crew (myself and Steve Rogers) at the back of the shop trying to direct a jet up into the fire. The door at the rear of the shop had fallen across the corridor so that its top rested just inside a storeroom opposite. We had been tasked to find the stairs. Everything was black but we could just make out thin strips of blackened wood of what was left of the stairs and so we did the best we could from the ground floor. I gingerly tested my footing and took position on the fallen door. The heat was intense.

Unknown to us (and we really should have been told and withdrawn), the decision was made to ventilate from the roof. Some tiles were removed, presumably by an aerial appliance. Heat and smoke raced up the staircase and out the opening. On the ground floor the relief was palpable. Now we had visibility and, looking left and right, were amazed to see the floor was as burnt through as the stairs, revealing a ten-foot drop to a cellar below. The door was somehow held in place by two beams beneath. I would hope good procedures would have made me realise even in such conditions that it is drummed into us to check the floor in front of you before placing any weight on it. But I couldn't help but view it as a potentially close shave.

The result of all those various incidents and experiences was that I at least felt like I had learned the ropes, so to speak. Yet cladding never arose as a topic. Nor had evacuation. Firefighters generally didn't get involved in such things. That was for staff to do before we arrived. High-rise procedures and stay-put seemed to work, incident after incident.

New beginnings

Around 1992 I had some personal problems that forced me to request a station nearer to home. I was very fortunate to find myself posted to Plumstead White Watch, a thoroughly professional team of firefighters under the management of a very old-school style Station Officer, Doug Thompson.

It was soon after this that another notch was raised on the temperature gauge for our frog in boiling water. The British Research Establishment, BRE, provided research, advice, training, testing, certification and standards to industry. Mock tests at the BRE identified a problem with 'Class 0' cladding panels such that they 'suffer extensive surface burning' often spreading nine metres to the top of the test building. Unfortunately, no changes were made to the guidance.[5]

The UK fire classification Class 0 was introduced into UK Building Regulations on fire safety in 1991. It is not a fire test but rather a classification from Approved Document B within those regulations. It is thus a Building Regulations rating and not a British Standard Classification.

It indicates the surface spread of flame and not the combustibility of the material itself when involved in a fire. Thus, combustible materials can achieve a Class 0 classification.

I have to say all this seems rather odd to me, reading about it after Grenfell. A test that measures the spread of flame across a surface but not the flammability of the interior? Back in the early '90s I had no knowledge of such matters at all. I only learned of the BRE when it was being privatised in 1997.

Concerning high-rise blocks, it would be useful to recall a couple of incidents I attended in my early years when I was at Plumstead. This was before the procedures became more rigid. We were called to a block close to the station. Having gained entry to a flat, we were confronted with a sofa on fire and the flat smoke logged from the ceiling down to perhaps waist height. The breathing-apparatus crew were standing by and two of us entered with an extinguisher to see if we could put the fire out quickly.

It was at this point that I received a message from our Station Officer, Doug, positioned on the ground floor outside the block. There was never any doubt from any of us in all the years we were there that he was in charge and knew what he was doing. His message was very simple, 'What have you got?' and 'Do I need to make it up?' By this he meant 'did he need to request more fire appliances?' Plumstead had two engines and we would have had eight to ten firefighters on that particular occasion.

At the same time the breathing-apparatus crew were asking if they needed to start up their sets. The two of us in the flat thought we could handle it with an extinguisher, so we told everyone to stand by. We had reasonable visibility and the air was breathable if we kept low. As it turned out, it took a little longer than expected and, a couple of minutes later, Doug came back asking for a decision. We requested one more minute as we used the last of the water in the extinguisher. A minute later he came back and said he needed an answer, or he'd request more appliances.

The point about this anecdote is twofold. First there was, and still is, a large element of trust. The Officer in Charge trusted the judgement of his crews and, standing outside and several floors below, was comfortable making decisions based on our word alone. Upstairs, we knew what he was asking and why. If we couldn't put it out with an extinguisher it meant breathing-apparatus crews, a 45mm jet and further appliances were needed. A decision needed to be made quickly and if there was any doubt we would withdraw, and he'd commit breathing-apparatus crews and request further appliances.

Secondly a certain level of discretion was allowed. We were able to assess the situation and decide whether we could take on a small level of risk by attacking a fire armed only with an extinguisher without breathing apparatus. The benefit was that we often extinguished small fires very quickly before they escalated. On the other side of the equation is the risk of placing two firefighters in an environment containing smoke without breathing apparatus.

Aside from the long-term effects of that, and all the other incidents, there is always the risk that a crew misjudges a situation. Because of this, the pressure on Fire Brigades generally has been to err on the side of safety. Thus, many years later the common practice was to set up a bridgehead two floors below the fire and only a crew in breathing apparatus armed with a 45mm jet was deemed appropriate to deal with a fire in a high-rise block.

As I recall we did manage to extinguish the sofa and withdrew immediately. Further appliances weren't requested, and the breathing-apparatus crew went in to ventilate, which simply means they cleared the smoke by opening the windows. Many years later this sort of scenario was rarer to the point of disappearing altogether. It was far more common in London to implement a much stricter procedure even for relatively small fires.

To be fair, looking back that was one incident where we certainly could have waited for the breathing-apparatus crew. There was no life risk and so I took in a small amount of smoke and gained what exactly? The fire was extinguished quicker and more efficiently (with an extinguisher rather than 45mm hose). On balance, very probably it wasn't worth it and so one can see why the Brigade moved towards implementing stricter procedures.

Another incident from this period concerns a more debatable scenario. Again, at Plumstead, in the early to mid-90s, we were called to a fire in a flat in a high-rise block. My memory today fails me when searching for the exact block. However, I think it was in Robert Street, a short distance from the station, and it was certainly during the day as it was light outside. On arrival we found a man apparently trapped on his balcony on about the sixth floor with heavy smoke coming from the flat. I was a designated breathing-apparatus wearer along with an experienced older hand, Kevin Richards.

We made entry into the flat armed with a 45mm jet. Our assumption was that the corridor in front of us led to the living room and the balcony was likely off that room or, if not, off a bedroom. We didn't know exactly where the fire was and visibility was poor, although one could see the blurred images of furniture a few feet ahead. The plan was to get the casualty first then extinguish the fire. Putting water on the fire would obviously create even more smoke, as well as steam, potentially making it worse for the casualty. As we worked our way down the corridor, we got a sense something wasn't quite right. None of the rooms showed any sign of fire and, from the end of the short corridor, we thought we had a view of most of the flat. Yet it was getting hotter.

We decided to act and this involved our first break with procedures. We split up. Kevin remained at the end of the corridor with the 45mm jet and I made for the patch of light a few yards ahead that I hoped was the balcony on which we'd seen the casualty. What we didn't know was, as the front door opened flush with the corridor wall, behind this now open door was a door to a bedroom. This was where the fire was located. In making a quick check behind the front door the closed bedroom door had been mistaken for part of the corridor wall. A quick sweep of the boot behind the door had failed to detect the architrave or door.

At the end of the corridor we were both unaware that the fire was behind us, a potentially dangerous situation for any firefighter. But casualties have priority and so I made my way the few yards across the living room, sweeping my foot across the floor in front of me as I did so, as well as sweeping my hand vertically up and down. This slows movement a little but prevents firefighters falling over furniture or being tangled in cables as we move through smoke.

I came into open air and found the casualty leaning over the balcony trying to avoid the smoke billowing around him. He was coughing and spluttering and in some distress. I looked down and was able to signal to the officer in charge (Doug again) that I had the casualty and got a thumbs up in return. It was at that point I noticed two things about the casualty. Firstly, he had a small dog in his arms, squirming with fear. Secondly, the gentleman, who was panicking almost as much as the dog, looked exactly like the character Terry McCann from the series *Minder*. Because of that, whenever I think back to this incident, I always think of the actor Dennis Waterman.

I quickly told our Dennis we would walk the few yards across the living room, down the corridor and out, reassuring him that everything would be okay, and I'd be with him the whole way. Now came our problem. He wouldn't go. It wasn't particularly hot, and the smoke wasn't especially thick or dark in colour. But, what to many firefighters is little more than a routine training exercise, is a scary situation for someone who has never been in a fire or experienced that level of smoke before.

No amount of coaxing or reassuring could change his mind. The smoke was getting thicker and I made it very clear he simply could not stay there. By now Kevin was shouting for an update from the other side of the living room. I could have waited for an aerial appliance to arrive but I wasn't sure we had the time, plus, if he was too scared to walk 20 yards through smoke, it might prove difficult getting him on to a ladder pitched 60 feet to his balcony. I considered forcibly manhandling him, but he was a little bigger than me and I judged this might prove difficult.

Now I would ask the reader to consider what you would do in this situation? If you remain there the man may not survive. Yet he simply refuses to go. A very frustrating situation when you know a smoke-free lobby and safety is just a few yards and seconds away.

The one thing you might consider, taking your own breathing-apparatus face-mask off and placing it on the casualty, places you in danger and also risks serious managerial discipline. Possibly the sack. The smoke was getting thicker, and I knew I had to make a decision. I looked over the balcony to check who was looking, stepped back and removed my helmet, making sure no one could see me from below. I depressed the first breath lever on the face-mask, shutting off the air until someone takes a deep breath. I placed the mask on Dennis's face and told him to breathe, getting him to hold it in place with one hand, the dog still squirming in his other arm.

I told him firmly I would lead him across the few yards across living room, down the corridor and out into the lobby where it was safe. Grabbing him by the scruff off the neck, I placed my other hand over the dog's eyes to stop it panicking and took a deep breath. With that I frog-marched him out of his own flat through the smoke and into the corridor, taking in a minimal amount of smoke as I did so.

What is the point of that story? It raises a number of important questions. What do we do when procedures do not cover every eventuality? What would I have done if the smoke was too thick and the distance too far for me to get myself out without my mask on? Should I have waited for an aerial appliance? But what if the balcony was beyond the reach of our tallest aerial? What if you knew it would take too long to arrive and set up? If the lobby and stairs had also been filled with smoke, how would I have got him and me down ten flights of thick black smoke? What would I have done if he had been one of the small proportion of people who are so terrified of putting a mask over their face that they simply refuse?

Did I do the right thing? I've no idea, even now. Maybe I had time to wait. The point is that you cannot calculate this accurately at the time. It's not a case of knowing for sure we have a 35 per cent chance of a good outcome if we chose A but B gives us 40 per cent. This is a point we will come back to with Grenfell. Many of the firefighters were placed in impossible positions with even worse choices. And no ability to calculate probabilities.

But this situation ended relatively well. The man and his dog survived. We re-entered, found and extinguished the fire and no one ever asked why the casualty came out with my mask on his face. If Doug, on the ground floor outside throughout the incident, ever knew I'd broken procedures he never said. If the fire was remembered at all by the firefighters, it was because of what happened when the man's wife turned up five minutes later whilst the poor man was sitting on the floor receiving oxygen. Instead of sympathy and concern, a whirlwind of rage and abuse came bellowing down the corridor and the poor husband could only sit there, with a blackened sooty face, and receive a wall of blame and anger.

She only paused her condemnation to enquire after the cat. The cat? We all looked at each other, then at the husband. He looked back. It was just at that moment Kevin walked out of the flat carrying a black bin liner with a grim look. He had held up the bag and was about to say something when he saw the woman for the first time. At the mention of a cat, he stopped short and looked at the bag and back at us. That look painted a thousand words. A brief pause ensued as everyone looked from the bag to the woman and the woman looked from us to the bag. And then the screaming began.

So into the annals of Plumstead White Watch history went the tale of when Kevin killed the cat. Of course, he had done no such thing, merely found its remains in the bedroom where it had sadly succumbed to smoke. Some recall the job as the one with the incredibly angry woman, but I always remember it as the job where we saved Dennis Waterman. I never did find out the gentleman's real name. What none of us ever considered was how we would have achieved that rescue if the lobby and the entire staircase were filled with thick, black, hot smoke. Back in the '90s it never occurred to any of us that such a scenario was possible.

Some people came out of the other flats and asked what was going on, as often happens at such incidents. Our answer was the same. There's been a fire in one of the flats but we have it under control. You are perfectly safe. Stay inside your flat. Stay put.

Around this same time I attended an incident in a high-rise block that emphasised just why deviating from stay-put might on occasions be potentially dangerous. It was on the neighbouring station's, Erith, ground and involved a quantity of rubbish left in the ground-floor lobby, including two mattresses. The only staircase opened into this outer lobby. Someone had deliberately set the rubbish alight and the only exit was full of thick black smoke. Had there been a communal fire alarm, scores of people all would have come down the only staircase.

If everyone had attempted to evacuate from the flats above one can imagine the possible scenario. Aside from the potential slips, trips, falls and crush injuries, a far more serious risk would be posed by the first person to exit the door at the bottom of the staircase. The thick black smoke, with nowhere else to go, would fill the stairway and engulf all the people coming down. The smoke would be hot and visibility would quickly be reduced to zero. Panic would ensue. Those at the bottom of the stairs might turn round to get back up and away from the smoke. Some might think making a dash down the last few flights and through the lobby might be a better option.

They would have no way of knowing the conditions in the lobby. Anyone coming out of their flats on upper floors at that precise moment and opening the door to the stairs would fill that floor's lobby with smoke. On that occasion none of that happened.

Residents stayed in their flats, which is what the building required them to do. The fire was extinguished and the rubbish cleared out. I know one argument will be that they should have a second staircase. I agree. But this does not help the thousands of blocks with only one escape route. We could fit every block with a communal alarm, but we must realise this is not a risk-free option.

I remember this incident well, even if the exact location escapes me. I had one other similar incident. Up to that point I had probably taken stay-put for granted. But I remember thinking it was lucky that the people did follow the procedures. Little did I know what was coming 20 years later.

Experts on stay-put

Prosser and Taylor lay out the arguments for the stay put policy in their book, *The Grenfell Tower Fire*.[6] They state the stay put policy has been the mainstay of fire safety in high-rise blocks for six decades and was seen as 'remarkably successful' until the Lakanal House fire in 2009. Since 1962 and publication of *The British Standard Code of Practice 3, chapter IV* the fire safety strategy for high-rise blocks has required residents in flats, other than where the fire started to remain in place. The fire is contained within the original compartment until the fire service arrives and extinguish it. Evidence demonstrated that this strategy was effective and avoided major casualties.

Importantly, the Code of Practice and other legislation and guidance made it clear that 'in an evacuation of a premise, no reliance should be placed on the fire service'. Interestingly, an expert witness to the inquiry stated quite clearly that the fire brigade were a fire and rescue service, not a fire and evacuation service. Another nuance much of the media seem incapable or unwilling to understand. Yet the reason should be obvious with a little thought. Firefighters are not present at the start of an incident. They arrive sometime later. They also arrive with only enough resources to deal with the incident.

For a stay-put strategy to work, residents and firefighters need to be informed and conversant with the policy and of what to do if the fire breaks out of that compartment. Active and passive fire-protection measures have been built into buildings to support the stay put policy and a 'defend in place' firefighting strategy. This is not just in high-rise premises. Hospitals and nursing homes are unable immediately to evacuate persons who are immobile. It would not only be impractical but dangerous to attempt to evacuate the very frail, vulnerable or those undergoing treatment or surgery.

In a high-rise, even if a lobby is compromised by smoke, the stairway should be smoke free. Other flats on the same floor are protected from the fire for one hour. This approach had worked and there were no losses greater than a single family in one flat in accidental fires in high-rise until 2009 with the Lakanal House fire. For decades evidence showed stay-put worked.

Having said that, it was recognised that one or more floors may have to be evacuated simultaneously and, after Lakanal, even a full evacuation. The problem was that how this was to be done was ill-defined. Many purpose-built high-rise buildings do not have smoke detectors or alarms in the common areas. In fact, they would be actively discouraged by fire safety officers as this would go against the stay-put policy. Without a central alarm system and no communication system, there is no way to inform all residents of the change from a stay-put policy to an evacuation. Additionally, buildings with a single staircase are not designed for large-scale evacuation.

The stay-put policy relied on a number of factors:

- The common parts of the building are constructed and maintained so that if a fire does occur in these parts it will be minor and will not spread;
- Any serious fire will occur within a flat;
- That flat will have a high degree of compartmentation;
- There is no reliance on external rescue.

Whilst stay-put was largely successful, these presumptions have since proved to be false. There have been fires in communal areas, especially where maintenance and enforcement have been poor. Fires do sometimes spread externally via balconies. Smoke can spread to other flats on the same corridor. Finally, compartmentation can break down. But the fact is we did not have those problems when I first joined the Brigade in the 1980s.

The main problem I encountered was residents not understanding the policy. On fire safety visits we would often have to explain to residents of blocks that in a fire in their flat they should get out and stay out. However, if the fire was elsewhere in the building they should stay put unless affected by fire or smoke. In that event again it is get out and stay out. I couldn't put an exact percentage on it, but a significant proportion of people did not understand stay-put. Which suggests that whoever managed the block was not proactive enough in telling people the rules.

Of course, stay-put only works if compartmentation holds. Compartmentation only works if someone doesn't bang holes through compartment walls and ceilings during refurbishment, doesn't neglect to maintain fire doors and doesn't cover buildings with solidified petrol.

Fire safety training and the British Research Establishment

We have already noted that firefighters are not fire safety officers. Around the mid-90s there was a pilot two-day training course for station staff to address what was seen by some as a fall in standards in firefighters' knowledge. It is worth noting why this pilot was quickly dropped. The reason we were given was that it was such a complex and technical subject it was felt they couldn't hope to train station staff to any reasonable standard in the time available. In short, a little knowledge could be a dangerous thing. Far better for firefighters to simply pass on queries about statutory fire safety than to feel over-confident and give poor advice. Whilst I would have preferred at least some meaningful fire safety training I can see the reasoning behind the decision.

It was around this time I recall first hearing about the British Research Establishment. This provided research, advice, training, testing, certification and standards to industry. The only reason I learnt of it was because it was around this that time fire safety officers were discussing the proposed privatisation. The general consensus among the fire safety officers was that it was a bad move and would allow commercial interests to influence decisions. Nevertheless, it was privatised in 1997 under the Conservative government.

With hindsight we can view this as another notch on the temperature for our frog in the boiling water of deteriorating standards. Previously, private building inspectors had been introduced and now the agency responsible for testing, certification and standards had been privatised. As noted previously, none of that necessarily meant that standards would fall automatically. So long as we kept high standards of testing and robust building controls there was no reason for alarm. What could possibly go wrong?

At this point regulations required rain-screen cladding to meet Class 0 classification but for insulation to be of *limited combustibility*. This rather technical term means a material with a specific density does not flame for more than 10 seconds and for which temperature does not rise by a specific amount.

On 1 May 1997 Labour swept to power with a 179-seat majority. The Grenfell Inquiry would raise serious concerns about BRE and testing and certification in general. Many of those questions arose during Labour's thirteen years in power. In 2022 the Fire Brigades Union passed a motion for its re-nationalisation. The union described the privatisation as a 'disastrous decision, opening the testing regime to commercial pressures and commercial interests', and said that re-nationalisation would 'ensure greater accountability, including a clear obligation to act in the public interest and without pressure from business and commercial interest'.[7]

An interesting incident

There's one incident from Plumstead that highlights the dilemmas firefighters can face. We had been called to a fire in a first-floor flat, persons reported. It looked like a semi-detached house had been converted into flats with an enclosed staircase built on the side leading to upstairs. On arrival in the early hours, we were told by multiple neighbours that a woman and her two children were in the flat. Smoke could be seen from the side and rear.

Two firefighters, Chris Millett and Roy King, donned breathing apparatus and gained entry, proceeding up the stairs with a charged jet. With everything done, I put a short extension ladder up to the first-floor front window to have a look. Through the gap in the curtains, I could see the room was free of any smoke or fire, but the door was closed. At this stage there was no point making a second entry; indeed, it could be dangerous, possibly creating a flow of uncontrolled air to the fire. But then the message came that Chris and Roy had somehow got stuck on the stairs.

I shouted to Doug asking if I should go in. We waited. The stairs had partially collapsed. With no time to waste I broke the window with a small axe. I did everything right, made sure my hand was above the pane as I hit downwards so the broken glass fell away from me. But then the latch stuck and I had to take my glove off to undo it. The gloves then were just too cumbersome. I ripped the back of my hand. Blood everywhere. I put my glove back on, felt blood trickle down the fingers and cursed my clumsiness.

But what to do? People outside were screaming about the kids. The Incident Commander was telling us the stairs are blocked. Going back to get breathing apparatus loses a good minute. Then we heard Chris and Roy were stuck. They couldn't get out to take over on the ladder. I called to Doug if we should go in: 'Go,' he said simply.

The window pushed up easily and I gained access. A colleague, Alex Marshall, followed me in with a hose reel. So, several procedures broken there: no breathing apparatus, no 45mm jet, and a gaping wound.

We did door procedure and entered the rear of the flat. To our surprise there was lots of fire but little smoke, the opposite of what one usually experiences. Visibility was fairly good. We both made a quick search. It was a small flat, just a couple of bedrooms. There was no sign of anyone. Had we missed them? Alex took another quick look in the kitchen and came back quickly: 'The gas taps were left on!' he said. What to do now? Everyone outside was so certain the family was home. We quickly decided to make one last quick search and get out. There was no sign of them. We got back to the front room and closed the door behind us, all the time half expecting an explosion. I've never come down a ladder so quickly.

It turned out the family weren't home. One lesson I've learned is that the public are sometimes wrong. I've sent a fire engine round to an alternative access at a job, having been assured 'you'd get a bus down there', only to discover it's a tiny alley a car wouldn't fit down. People panic, get excited, eager to help or just make honest mistakes. The fire, we learnt later, was suspected arson. Hence the multiple seats of fire we saw on entering. He'd left the gas taps on the cooker, but I guess without pushing them in it didn't work as planned. Or perhaps we were just lucky. I ended up getting stitches and still have the scar today on the back of my hand.

What does this demonstrate besides that you should always keep your gloves on? Sometimes firefighters are placed in a situation where they have to make a quick decision. You can look back with hindsight and think perhaps some of those decisions might have been wrong. Perhaps there was a better option. An investigation might highlight procedural errors and you might well find yourself under some managerial action.

But when people are screaming at you and you think a mother and two children are involved it is easy to make a snap decision. Sometimes potential outcomes outweigh the risk and warrant bending procedures. Back in the '90s I felt I could at least argue the case for my decision. There were no signs or symptoms of a back-draught or flash-over so a hose reel was acceptable. There was potential to save life and time was a factor, thus an attempted snatch rescue was within the boundaries of reasonableness. The problem as I see it is that you are very unlikely to win such an argument today, even in a scenario where most firefighters and the public would expect such action. Unless, of course, it was large incident with lots of media attention.

I had another incident with Alex. This time we were in breathing apparatus in a top-floor flat above a shop. The building itself was unused and partially derelict, potentially used by the homeless. Keeping to the edge of the room we had managed to put out all visible flames on the far side. Visibility was still very poor with lots of thick black smoke. We got the okay to ventilate and I was about to open the window when Alex stopped me. Something didn't feel right. He shuffled forwards and found a large hole in the middle of the floor. We decided to delay ventilating and investigate. Good job we did.

The fire had started in a sofa on the top floor. It must have been bottled up for hours but with just enough oxygen to maintain itself. Burning through the sofa and floorboard underneath, the remains of the sofa were now in the room below, still alight. Opening the window would have created a chimney affect with us standing between the fire and window opening. Such are the potential dangers of firefighting.

Garnock Court and boiling frogs

In 1999 a cladding fire occurred at Garnock Court, Irvine, Scotland. Fire spread from fifth floor to roof. A subsequent report by Dr Moore, an expert from the Fire Safety Development Group, to a parliamentary committee explained the difference between limited combustibility and Class 0 tests.

A material could be combustible and still achieve a Class 0 rating by adding fire-retardant chemicals or facing the combustible material with a metal foil or sheet: 'this serves to undermine the integrity of the regulations.' MPs advised ministers to scrap Class 0 standard and require cladding systems to be entirely non-combustible or to pass a large-scale test at BRE.[8] The following year the government did not follow the advice and instead retained Class 0 standard and introduced large-scale tests at BRE as an alternative.

In 2001 at the BRE a test of a cladding system consisting of non-combustible glass-wool insulation and ACM cladding panel failed disastrously. Flames extended 20 metres, twice the height of the nine-metre rig, in less than six minutes. This was despite ACM being of Class 0 and supposedly being able to be used on tall buildings in compliance with guidance in Approved Document B. A subsequent report from BRE to government the following year did not make clear the seriousness of the situation.[9] In 2002 the CWCT, an organisation representing the cladding industry, warned that new testing methodology which resulted in a number of fails could result in abandonment of use of rain-screen cladding with economic consequences for the industry.[10]

If anyone expected the New Labour government of 1997 to change direction, the above details should demonstrate that this was a vain hope. Back at station in London I heard nothing about Garnock Court. I don't recall any emergency information warning us about cladding fires. There was certainly no practical training exercise. Equally nothing regarding evacuation. I had never heard of 'rain-screen cladding' and was blissfully ignorant of building regulations.

New station

In 1999 I was posted to Orpington Red Watch as a Sub Officer. Orpington was a one-appliance station. Community Fire Safety work had been an increasing part of our work. This is distinct from statutory fire safety. In essence, it involved giving fire safety advice to the public and fitting smoke alarms. Much of this work was done in residential high-rises. As such we became very familiar with the standard of maintenance in such blocks. At the time I wouldn't have claimed all blocks were kept to a high standard. But looking back the impression I get is there was a marked drop in standards after my time in Orpington. We will come to the possible reasons why shortly.

Now being in charge of a watch I came into contact with statutory fire safety more often. One of our jobs was what was called 'G' visits. These were jobs delegated down by fire safety. One such G visit was to inspect a high-rise. This involved checking that the fire lift worked, the dry riser inlet was accessible and operational and that each outlet on the floors also worked. Along the way we'd also check all the fire doors and that the exit was clear. Any defects would be sent to fire safety and they would pass it on to whoever was responsible for the block, usually the local council. A few weeks later we'd get a revisit job to check that the work had been completed. This was usually triggered by the council after the defects had been rectified. It did not go unnoticed that sometimes this took weeks. At the time I thought this was poor, given the fire safety of the entire block relied heavily on such matters. How little did I know just how low standards were to drop.

This is probably a good point at which to deviate into equipment and procedures.

Fire engines and equipment

Basic fire engines come in three main types. All three are the same vehicle but with slightly different equipment. A two-appliance station has a Pump and a Pump Ladder. The main difference being that a Pump carries a 9-metre ladder (reaching a second-floor window) whilst a Pump Ladder carries the larger 135-ladder (reaching a third-floor window). A one-appliance station has a Dual Purpose Appliance, DPL, which carries both a 135- and a 9-metre ladder.

The distinctions between these are further complicated by occasionally all being referred to as 'Pumps'. So a ten-pump fire is a fire where ten pumping appliances attended although they might have been a combination of all three types of appliance.

Each of these fire appliances has an internal tank of water, 300 gallons in all, around 1,365 litres. The appliance has two hose reels, one on each side coiled around a drum in the middle lockers. The hose-reel works straight from the water tank without any need for setting into a hydrant. It can supply 115 litres per minute. Thus, if used constantly, a fire engine could supply one hose-reel with water for nearly 12 minutes. In practice, one might use the water a little more sparingly.

The fire appliance also carries five lengths of 70mm hose and two lengths of 45mm hose on each side, ten and four lengths in total. The hose between the hydrant and the fire engine is always 70mm. A jet, the hose used to attack the fire, can be 45mm or 70mm. The former is more manageable and the one recommended for compartment fires. Thus a 45mm jet consists of a number of lengths of 45mm hose with a hand-controlled branch attached to deliver the water.

A 45mm jet supplies 450 litres per minute and so we can see that a tank of water will only last about three minutes if a 45mm jet was used. Two jets and it's just 90 seconds. For this reason, it is imperative that a hydrant is located and got to work as quickly as possible. In the past, knowledge of one's ground and a physical 'hydrant book' was the only way to locate hydrants quickly. In London we are blessed to have a hydrant on most roads. Rural areas are not so lucky.

Each appliance has five breathing-apparatus sets. Each set will supply air for about 31 minutes if breathing at a normal rate. This is more than enough for most domestic fires, including those in high-rise blocks. Hard work reduces this significantly. Extended-duration breathing-apparatus sets carried on Fire Rescue Units last for 47 minutes.

In addition, appliances carry a variety of other equipment: breaking-in gear, lines, general tools, air bags, cutting and spreading tools, disc grinder, reciprocating saw, lightweight portable pump, portable generator and lighting, and many others.

Domestic house fire procedure

At a two-appliance station the jobs at an incident are assigned at roll call at the start of the shift. The two firefighters on the rear of the Pump Ladder are the breathing-apparatus crew and the Pump's crew are detailed to get the hydrant sorted and roll out a jet. One driver will act as pump operator whilst the second will be entry control officer for the breathing-apparatus crew. A one-appliance station should have a crew of five and so you have just enough for a pump operator, entry control officer, two breathing-apparatus wearers and the officer in charge.

Just to illustrate how this works, on one of my last working jobs at Addington we attended a house fire which turned out to be a garage converted into a flat attached to a larger house. As an aside, to show how fragmented our fire safety legislation has become, I became aware of a similar property when fitting smoke alarms. The garage had been converted into a downstairs bedroom but the garage door had been left in situ, presumably to disguise the conversion. The only exit from this room was via the kitchen. This meant any fire in the kitchen would cut off the occupant. When I raised this with fire safety I was advised to contact local building control. When I did so, they said that if it was more than two years old there wasn't much they could do.

This seems a ridiculous and dangerous situation. Either something breaks building and fire safety regulations or it doesn't. How many other similar practices are simply allowed to happen or are not reported at all?

Back at our incident I had arrived early for a day shift and let the Blue Watch Officer in Charge go before the official end of the shift. When the bells went down, I had a mixed crew of Blue Watch and my own Green Watch who had jumped on for their Blue Watch counterparts. It was one of those jobs where everything went exactly like clockwork despite there being mixed crews. I nominated one appliance as the pumping appliance and the second as the Incident Command Pump, from which messages are sent and received to control. The Ladder's crew donned breathing apparatus while the Pump's crew set into the hydrant, set up Breathing-Apparatus Entry Control and laid out a hose-reel and covering jet.

Initially there was the possibility that there were persons involved, as a mother and her two children lived there. Based on that, I directed the breathing-apparatus crew to enter with the hose-reel rather than wait for the 45mm jet which was still being prepared. A decision that might get a black mark from an observer at a training exercise, the mantra now being 'a compartment fire requires a 45mm jet'.

In this situation we had three potential rescues, a smoke-filled extension and two breathing-apparatus wearers ready to go with a hose reel. With a 45mm jet being laid out that would take another minute or so to finish and fill with water. If there had been signs of back-draught or flash-over I might have made them wait. But it seemed a calculated risk and still seems reasonable looking back. The upshot of this was that the crew entered the property within a minute of us pulling up.

Or however many seconds it took for them to get off and start their sets up. Incidentally, this is another calculated risk. You are not supposed to don your breathing apparatus en route in case of injury. But with a potential 'persons reported' firefighters often do just that so that they save a few vital seconds when the appliance stops. You can even depress the first breath lever (cutting off air supply to mask) and turn the set on so that you are ready to go. So the appliance came to a halt and the breathing-apparatus crew just had to remove helmets, don face masks, pull fire hoods up over heads and put helmets back on.

By the time they had done that, one of the pumps' crew had pulled off enough hose reel and handed them the branch at the door. Another firefighter made entry, with a sledgehammer if I recall. As they searched, they quickly found the small fire and extinguished it. Meanwhile the jet was made ready, another covering jet laid out, casualty handling area set up and messages sent to control. Enough was done to free up two firefighters to stand by as an emergency crew with breathing apparatus just in case.

With the 45mm jet ready we were able to swap over the hose-reel jet. Worth pointing out that many a Watch Manager has had his ear pulled by an oncoming Station Manager about why the hose-reel was used at all. Now that I'm retired I can say that sometimes the hose-reel is the better option. It's quick, it's ready to go and if there's a rescue and you are the first appliance on it's the best option most of the time. To see why, one only has to consider an incident when there's only one appliance available.

Before we do it's worth commenting on the above incident. In my opinion that was an incident I can look back on as going well. It was little to do with me being in command; it was mostly down to the firefighters all being very competent, knowing their jobs and performing to a high standard.

In the end the woman was not home, being on the school run with her two primary-aged children. I found out later that the council had some trouble finding alternative housing for her (if I recall she was receiving housing benefit and paying rent). In the end (a few days later) a place was found a few miles away from the children's school. So one council took several days to find a place for one person. She wasn't happy with the extra travel but had no other options. It is worth remembering this when wondering why a council has trouble finding scores of places at once.

Let us now turn from a two-appliance job that went well with adequate resources available and compare it to one where only one appliance is available for a considerable time.

One-appliance job

Quite a lot of stations are one-appliance, meaning that they only have one fire engine and a minimum crew of five firefighters. Those stations with two fire engines have minimum crews of four and four. Here's an example of an incident I attended when my second appliance was elsewhere. I was riding in charge of the Pump Ladder with a crew of four. This should demonstrate the difficulty in trying to apply 'safe systems of work' in the initial stages of a job. We'd been called to a house fire, 'persons reported'. On arrival, we were confronted with a semi-detached house with a man hanging out of the upstairs window and a significant fire in the downstairs kitchen.

The priority is, of course, the rescue. Two very experienced firefighters, Dusty Taylor and Glenn Higgins, were riding on the back and donned breathing apparatus. But by rights they can't make entry without a Breathing-Apparatus Entry Control Officer or without water. Pete Forster, the driver, was the pump operator and busy getting the pump to work. He was not supposed to be pump operator *and* entry control officer *and* set in the hydrant. I was supposed to stand back and monitor things and not get involved in entry control or pumping. But we can hardly stand by and tell the guy he will have to wait to be rescued until another appliance arrives.

So here is what happened. I grabbed the short extension ladder, Pete engaged the pump and got the hydrant gear while Glenn and Dusty put their breathing-apparatus sets on and started to roll hose out to prepare a jet. If I recall on this occasion it was decided that a hose-reel was not sufficient. There was a lot of thick black smoke and the fire seemed to be well established. Pete was looking rather flustered, trying to carry the entry control board in one hand whilst getting the hydrant gear off the back of the appliance. He shouted something about what I wanted him to do first, so I grabbed the entry control board and suggested pump first, then hydrant. We'd sort the jet out.

Glenn and Dusty hadn't started their sets up yet, though they had them on their backs. They shouted over if I needed help with the ladder. No, I told them to get the jet ready while I got the ladder. Organised chaos is a phrase often used. It might seem chaotic to a bystander but, in reality, it was about prioritising quickly.

So Pete put the pump in and went to find a hydrant. Not an insignificant job when you are on your own and the hydrant is two or three lengths away, meaning four to six lengths of hose need to be rolled out and connected. Dusty and Glenn started preparing the jet whilst I grabbed the ladder, having dumped the entry control board against the garden wall. All this time I'm shouting at the guy to hold on and not jump. Whilst he's only one floor up he could easily break an ankle.

The short extension ladder is a three-piece ladder which is fairly easy to put up on your own. I managed to do this quite easily by which time Dusty and Glenn had already got their jet ready and turned on the water. Normally this would be the pump operator's job, but Pete was busy rolling out hose from the hydrant. They shouted over to ask if they should gain entry but I told them to hold on until I had got the man down.

As they grabbed the enforcer tool to force entry and started up their air, I quickly climbed the short distance and helped the man onto and down the ladder. Another break with procedure here; I didn't get anyone to foot the ladder. But, of course, there wasn't anyone spare and I only had to climb three or four rungs to reach the gentleman. So, the man was brought down and he sat on the fire engine where he told me anxiously that his dog was still in the house. The breathing-apparatus crew forced entry and I quickly grabbed their tallies and put them in the board. Another break with procedure, no Entry Control Officer. As they went in, I shouted after them about the dog.

At some point a second appliance arrived and I was able to put someone on entry control, someone on the casualty, get a second breathing-apparatus crew to stand by and get the officer in charge of that appliance to send a message. A short time later Dusty came out with the dog in his arms safe and sound. The fire was extinguished and damage was confined to the kitchen.

Now we could have done that differently. I could have ignored the fire and got my crew of three to pitch the ladder and rescue the man whilst monitoring. Then put someone on the casualty and prepared the hydrant and jet and wait for the oncoming appliance before having enough personnel to commit a breathing-apparatus crew.

However, this would lose some of the benefits from a quicker entry. The fire damage to the man's house was confined to the kitchen and we rescued his dog. I think most of the public would agree firefighters take on a small amount of risk in a calculated way to mitigate damage.

The point of this tale is to show the difficulties in the initial stages of an incident and the decisions that have to be made. No doubt I have made mistakes in my career, big and small. But I don't think so in this case. The problem is of course that I have no prior knowledge of the probabilities involved? What are the chances of a back-draught or flash-over? What are the chances of a gas explosion and the house collapsing with a crew inside? What are the chances of the entire house being destroyed if we delay entry by five minutes? or two? or ten? When exactly will the second appliance arrive? What are the chances that the man forgot or wasn't aware that his son hadn't gone to the shops and was actually cowering in a rear bedroom?

All I can do is react to what is in front of me. A rescue. A kitchen fire. A fire large enough to warrant a jet rather than a hose-reel. A pet claimed to be in the house. A limited number of personnel. I think on balance we did as well as we could with the resources we had. This didn't prevent an officer being sent to the station the following day to give me a telling off. And I shall relate it here because it's interesting how organisations sometimes behave.

The telling off wasn't for any of the corners I described cutting above. By the time anyone senior arrived we had everything in place: an Entry Control Officer, casualty-handling area, covering jet laid out, etc. Instead, in my message I had indicated we used the 'wrong' ladder. Apparently, someone who had seen my incident message had decided the short extension ladder was not considered a rescue ladder and I was to take this on board in future.

It is worth noting that this incident was after I had spent nine years in training. I pointed out that we had been teaching trainees that the three-piece extension ladder was indeed the best ladder for the first floor, the 9-metre ladder for the second and 135-metre for the third or, if you are lucky, the fourth. We certainly hadn't been teaching that the short extension could not be used for rescues. Nevertheless, he insisted that this was now wrong. The Fire Brigade is a disciplined service and, being more of a rugby than football man, I should really encourage people to accept what a senior officer, or referee, decides.

However, there are times when even quite small things are important. The point here is that a 135- or 9-metre ladder simply would not fit under the window on the first floor. It would be banging into the wall above. We then got into a discussion about the fact that a ladder should be placed a third of the working height from the building. If you can forgive me for being a little technical, basically this means either not too shallow so that it might slip from under you and not too steep making it difficult to climb or descend. So, it dissolved into an argument about whether a 9-metre ladder could fit under a first-floor window. He maintained it could. I maintained there'd be plenty of times it couldn't.

Unfortunately for the officer, I had previously been a maths teacher and, at the words 'Pythagoras' and 'hypotenuse', the blood drained from his face. The poor man had been sent to berate some junior officer over a minor quibble about the contents of a message, only to find himself in a maths lesson he no doubt wished he'd seen the last of thirty years before.

Out came a calculator, paper and pencil and I was able to prove, using Pythagoras' theorem, that a ladder 5.3-metres long before it had even been extended could not be positioned safely under a window often just over 3 metres from the ground. I had never seen a man back out of a room so quickly. Of course, the killer argument had nothing to do with a long dead Greek mathematician or philosopher at all. The fact was we didn't have a 9-metre ladder. Pump ladders on two-appliance stations only carry 135 and short-extension ladders. Turned out the officer had merely been sent down by someone higher up who didn't like the contents of the message and, for reasons only known to them, believed short-extensions ladders were not appropriate for rescues. Not just mission but assumption creep.

The point of this is not to crow about a minor petty victory over bureaucracy from years ago, as satisfying as that can be. The point is organisations evolve and grow to the point that they sometimes forget what their core purpose is. I am pretty sure the only thing the man rescued was interested in was how long we took to get there, the fact that we rescued him and his dog and how much of his house we saved.

Having seen the difficulties of having a single appliance at a house fire it should be obvious how much more difficult it is at a high rise. Hence why the pre-determined attendance at a residential high-rise has always been multiple appliances. At the time of Grenfell, it was four. Let us now turn to high-rise procedure in general.

High-rise procedure

The first thing to say about high-rise procedure is that it is focused on the initial fire compartment. The priority is always casualties, but it is expected that they will usually be in the flat where the fire started. It is also expected that compartmentation will remain intact. By this we mean that each individual flat should contain the fire for a minimum amount of time.

It is this concept upon which the stay-put policy is built on. This section concerns the high-rise procedure that evolved over the years and that was used by the London Fire Brigade at Grenfell.

Let us imagine a two-appliance station called to a fire at a high-rise residential block. Two fire engines will arrive, perhaps five minutes after the initial call. This arrival time is affected by the distance, time of day, traffic and many other variables. If the closest station is already on a call, then the next available station may be another five or more minutes away. In addition to this is the fact that the initial call itself may have been some time after the fire started. I've attended fires that have clearly been 'bottled up' for a considerable time and we only got the call when a neighbour or passer-by thought they smelled smoke.

It is precisely because the arrival time cannot be guaranteed that it was explicitly stated to the inquiry by an expert that the building should not rely on the Fire Brigade for evacuation. Again, this does not mean the Fire Brigade won't assist an evacuation but that the system in place should rely on something other than the Fire Brigade. The initial attendance arrives with enough resources to deal with the incident. There are no spare firefighters to engage in evacuation. There is no 'evacuation team' standing by.

Back to our scenario: our two appliances are called a couple of minutes after the fire started and arrive within five minutes, similar to Grenfell. The building is divided into sectors. The 'Fire Sector' includes the floor the fire is located on plus the floors directly above and below. Above this three-floor fire sector is the 'search sector'. Two floors below the fire is the bridgehead. This is where firefighters set up a type of forward command post from which to fight the fire.

Equipment and personnel are brought up to the bridgehead. Often this is done via the lift. A 'firefighter's switch' on the lift is activated making it come to the ground floor so that firefighters can take control. The buttons on the floor lobbies are de-activated and the lift can only be controlled by a firefighter in the lift car. Below the bridgehead is the 'lobby sector'.

These sectors can be seen in figure one. Figure 2 shows how the crews of two appliances might be deployed: four firefighters upstairs and four on the ground. Those on the ground have to find a hydrant and connect hose from that to the fire appliance. They then run hose from the fire appliance to the dry-riser inlet at the base of the tower. The fire appliance acts as a booster pump sending water up the dry riser, basically a vertical pipe.

At the bridgehead a breathing-apparatus crew starts under air and connects hose to the nearest dry riser outlet, preferably the floor below the fire floor. As soon as two firefighters are available, or another appliance arrives, a back-up breathing-apparatus crew should assist the first crew attacking the fire. Whereas the first crew sets in their hose to the floor below, the second crew sets in on the floor of the fire.

All this looks very neat and tidy. In the unlikely event of fire or smoke spreading to the floor or adjacent flat it would not take much to extend operations to counter that. It would never have occurred to any of us in my

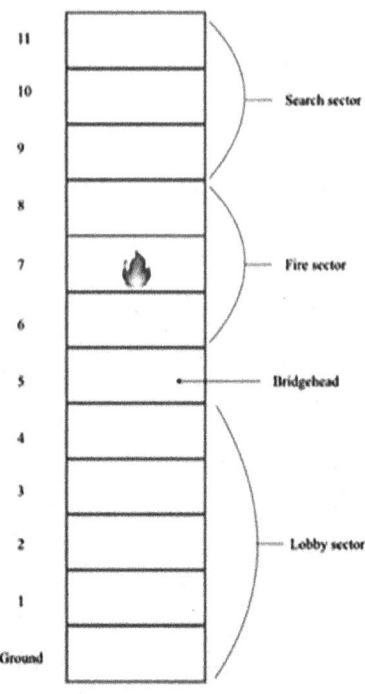

Figure 1: High-rise procedures: Sectors.

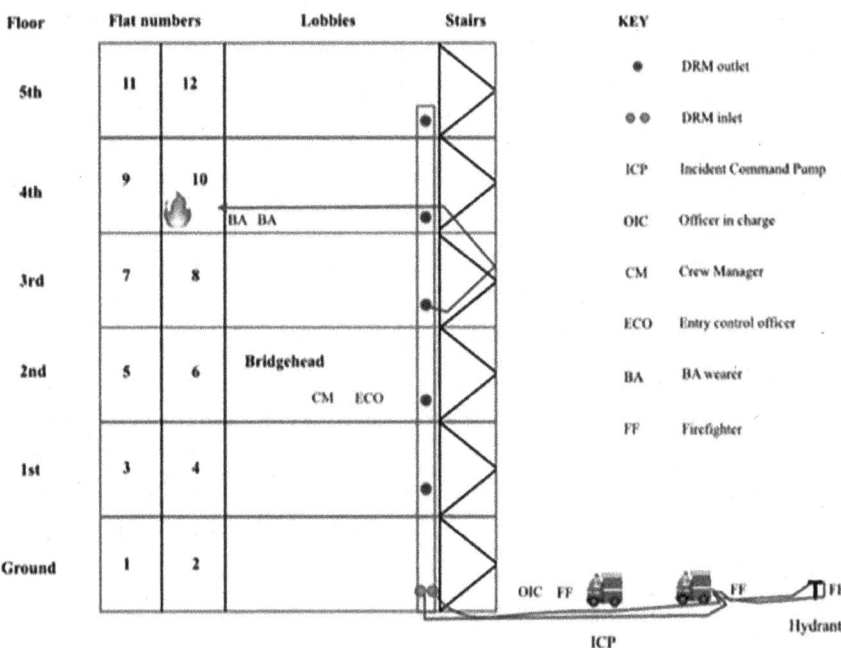

Figure 2: High-rise procedure.

early career that one might get multiple fires in multiple flats at the same time, outside of an arson situation.

The only concern I heard was about the insistence that the first crew ran their jet from the floor below the fire. This meant running hose out of the door on that floor, up the stairs and through the door on the fire floor, leaving two doors ajar and clogging up the stairs. Many felt this went completely against previous good practice. Firstly, all doors are supposed to be kept shut and, secondly, the often only staircase was supposed to be kept free of obstruction.

An alternative would involve hand-controlled dividing breeches. This is a Y-shaped connection that could be attached to the dry-riser outlet on the fire floor. From there, two lines of hose could be attached. A lever on each arm of the 'Y' could control the water to each jet. Thus, you could have one jet go to work and a second one attached later.

I have no idea why this was not considered. There are buildings where dry-riser outlets are only on every other floor (another building practice that seems ludicrous) rather than on each floor. In those cases, it cannot be helped. But, in general, many felt it was a regressive step to have a policy that caused fire doors to be left ajar and have hose on the only stairs to the exit. Looking back, I haven't changed my mind. Nevertheless, those are the procedures that had evolved up until the tragedy at Grenfell. Hopefully, the reader now has a clear idea of what the initial crews were attempting to do.

Unlike commercial buildings the occupants of a residential high-rise must rely on their own knowledge of a building's layout and fire procedures.[11] For this to be the case they have to be aware of what that policy is. In the case of residential high-rise buildings, this invariably meant the stay-put policy.

Much of the tightening up of procedures came from severe fires. Two examples resulted in firefighter fatalities. A fire on the fourteenth floor of Harrow Court, Stevenage, Hertfordshire in 2005 resulted in the deaths of two firefighters. Having rescued one person, they returned for a second casualty and were caught in a flash-over or back-draught. One attempting to escape the compartment had become entangled in cables that had fallen across the door when the trunking holding the cables had melted. In 2010 a fire occurred on the ninth floor of Shirley Towers, Southampton. This, too, resulted in the deaths of two firefighters. The flats were a complex 'scissor' design over two floors which contributed to some confusion.

The recommendations from these fires are far too widespread and complex to cover here. However, one important and relevant point was that crews should extinguish fires before proceeding past or above the fire.[12] They should also have a charged jet. If attacking a fire they should have a back-up crew as soon as possible. Such were the procedures at residential high-rise blocks by this time in my career.

One important concept to note: a fire survival guidance call is not confined to high-rise blocks. However, several dozen occurred at Grenfell that it is worth covering at this point. A fire survival guidance call, FSG, is a call received by control, via the 999 system, where the caller believes that they are unable to leave their premises because of fire, heat or smoke and the operator remains on the line. The policy reiterates that the advice is that it is usually safest to remain in the property unless affected by fire, heat or smoke. However, callers will be advised to leave their property if they start to become affected. Only cases where they cannot leave are classed as FSGs.

In these cases, information as to number of people, location, proximity of fire and conditions will be passed on to the incident commander. At smaller incidents, or the initial stages of larger incidents, this will be done via the incident command pump. This is simply a fire appliance designated as such and manned by a firefighter to relay messages between control and the incident commander. At larger or protracted incidents, a command unit will perform this function.

Incidents of note

I recall a number of interesting incidents from my time at Orpington. A road traffic collision on the M25 with a car going under the rear of a lorry, the driver's foot trapped under the pedals. Frantic efforts to cut him out with the paramedics warning us that time was critical and that they might have to amputate his foot on scene. Then, just as the decision was made to do so, somehow, miraculously, his foot eased out. I've no idea if he realised how close he came to losing his foot.

An elderly gentleman in a dilapidated flat where the breathing-apparatus crew struggled to get him down the corridor and out of the door. Up until recently I still had the details in an old notebook I'd left discarded in a drawer; address in St Paul's Cray, flat of five rooms etc. I remember feeling saddened at the state he was living in as I got the impression that he wasn't looking after himself very well. Then the quietness round the mess table when we learned he was a retired firefighter. An incident where a small oxygen cylinder exploded in a shed just as Trevor Lee poked his head over the wall a few feet away.

A double fatality at a head-on road traffic collision involving a young woman and middle-aged man. A mobile phone going off in the car as we tried to cut away the crumpled metal. A paramedic sadly shaking his head and telling us there was no point rushing now. One of them must have simply taken their eyes off the road for a second and drifted a little too far on the bend. Such is the precariousness of life. Look to change channels on the radio or get a sweet out of your bag. A split-second break of concentration.

A minor road traffic collision in a country lane near Chelsfield. A police car arrived and I was amazed to learn that they were the only police car on duty that night in the whole of the Borough of Bromley, with a population of over 300,000 and the largest by area in London. A point I would ask the reader to remember when we touch on building control officer and fire safety officer posts being cut to the bone.

Another road traffic collision where a car hit a horse which sadly had to be put down as we struggled to cut the driver free. An air ambulance landing a little too close and the wind from the rotors covering us in glass and debris. Having to quickly get a sheet over the poor casualty to protect him.

Perhaps the most memorable was the following. It was a sunny Sunday afternoon around the year 2000 and Liverpool and Manchester City were playing out a particularly dreadful 0-0 draw on the television. An elderly gentleman, old enough to have fought in the war (so he told us after), grew increasingly bored with the match and had gone to the kitchen to put something on the hob for lunch. Then he returned to the living room and sat back down in the forlorn hope of a goal. Ten minutes later he gave up and, forgetting about what was cooking, went upstairs to have a bath.

At this point of the tale, we learned he had a prosthetic leg, which he removed to get into the bath. A little while later he smelt something, got out of the bath and, leaving his prosthetic leg behind, hopped downstairs. Downstairs the dismal game was coming to an end as he hopped across his front room stark naked and dripping wet. He quickly realised his mistake. Whatever he had left on the stove had caught light and the kitchen was ablaze.

He made his way back across the room and, getting to the front door, made what could have been a fateful decision. Do you escape from the house stark naked or nip upstairs to throw something on or grab your keys and wallet and even, in this man's case, a false leg? The Brigade would advise the former: 'get out and stay out'. This man chose the latter. By the time he'd reached the top of the stairs the smoke had overtaken him. Deciding to leave his leg in the bathroom he dived into his front bedroom and shut the door behind him. There, fortunately, he was safe for a while, at least from the smoke. Luckily, he had a phone in his bedroom.

At the station the bells went down, the teleprinter whirred and the duty person handed me the slip as we jumped on the motor. It read a fire in a house with one person trapped in the bedroom. On arrival we were confronted with the elderly gentleman sitting on the windowsill, still stark naked with his one leg hanging down over the edge.

A breathing-apparatus crew was ordered to get ready whilst another crew pitched a ladder and a firefighter, not in breathing apparatus, went up to assist the man. To get the man onto the ladder safely, he had to get back into his bedroom and the firefighter had to enter the room and assist him from behind whilst a second firefighter, also not in breathing apparatus, went up to help from the front. All this time a small amount of smoke was swirling out of the window, but the bedroom door thankfully remained closed.

The incident also stuck in my mind because we made a mistake. As I was in charge at the time it would have been down to me. I had told the breathing-apparatus crew to make entry but not to go upstairs. Instead, they were to head for the fire in the kitchen and extinguish that whilst the rescue was in progress. I thought I'd made it very clear not to go upstairs and open the bedroom door, the reason being that I did not want the casualty or the firefighters to be exposed to smoke. They were in a relatively safe space with only a minimal amount of smoke.

Of course, the reader will already guess what happened. Which raises an interesting point. I can be certain my message was clear. The crew can be equally certain what they heard. But there are times in the heat of the moment when the intended message just does not register. I can only guess my insistence not to go upstairs was heard as a direction to do just that.

The result of this misunderstanding was the breathing-apparatus crew went straight upstairs and opened the door to the bedroom. At this point the elderly gentleman was back sitting on the windowsill with a dressing gown flapping uselessly round him. The firefighter on the ladder was trying to get him to turn round and assist him down backwards. The second firefighter was still inside trying to help from the bedroom. When the crew opened the bedroom door all three were immediately engulfed in thick black smoke.

Fortunately, the gentleman was spritely enough to quickly hop down the ladder facing the wrong way, with his back to the house, assisted by the firefighter on the ladder. The firefighter in the room made a quick exit as soon as the man was clear. The gentleman had no ill-effects, and he was quickly re-united with his prosthetic leg.

Another interesting type of incident involved hoarders. When I first joined if we dealt with a hoarder that was the end of it. Now we would notify the local council, possibly social services, and place it on the operational risk register, which could be accessed via the appliance mobile data system. Of course, we had no powers of entry to a domestic dwelling, but it was useful to have prior knowledge that you were attending an incident at premises affected by hoarding.

My first experience of a hoarder included dozens of cats, dogs and caged birds. I had been part of the first crew in and was aware of some of the cages but with visibility so poor it was difficult to make out what exactly was impeding our progress. It was only after the fire was out and we attempted to re-enter without breathing apparatus that we realised what the smell was. The tiny flat was covered with the remains of dead animals and not all from the fire. Animal faeces was everywhere, even the bath. Such a terrible sight and smell that it still makes me feel slightly sick even today.

The saddest was perhaps one very close to Southwark fire station. The breathing-apparatus crew were in the corridor of a flat, having difficulty in gaining access to the rooms. Unknown to us, this was because each room was full of black bin liners, each full of decades of items. The fire appeared to be in the last room, and it was a big fellow called Bob Smith who pushed the door so hard that it snapped in two diagonally from about halfway up to its top corner. This pitched him forward into the room on top of a layer of bin bags.

Just as well. A small back-draught rolled out of the room and across the corridor ceiling, roughly from about head height where Bob had been standing. I was on my belly feeding the hose reel in and, not being in breathing apparatus, decided discretion was the better part of valour. The smoke level gave about a foot's worth of clear air, so I turned and crawled out. Once the fire was out, we had to clear the flat to check the fire had not spread behind the floor or skirting boards. The result was a huge pile of burnt material on the grass outside.

The sad part came the next morning when we realised that we had lost a piece of equipment, a ceiling hook (a long stick with a point and a bill on the end for tearing down ceiling plaster). Returning quickly, we found the item sticking proudly like a flag atop the mound. Beneath it the elderly lady had a roll of new bin liners and was carefully filling each one with the burnt remains of her flat.

One last example of a hoarder occurred in a rather nice Edwardian house with one side obscured by a huge tree. On entering, a 'ramp' of rubbish led from the front door, down the corridor and into the kitchen. At the cooker the rubbish was about three-feet high with a small trench dug around the cooker to access the controls. What was amazing was the front room. Enough to take one's breath away. A sea of green presented itself. The tree outside had grown through the window and branches enveloped the ceiling and walls so that one could hardly see them at all. It was like a scene from *Life on Earth*. I remember the poor woman outside, embarrassed and tearful.

Summary

Hopefully this first part of the book has given the reader a flavour of what life as a firefighter entails, the types of incidents, procedures and basic concepts. It should also demonstrate the dilemmas firefighters sometimes face. The constant balance of risk and reward. At the same time, management slowly moving goalposts towards the risk averse. Sometimes rightly, no doubt for good reasons but with a great danger of forgetting what the public expect us to be there for.

Of all the various incidents, those at high-rise blocks were among the most frequent. They were also the most trained for. And yet, in those first seventeen years of my career, I don't recall anyone ever questioning stay-put. I never heard of rain-screen cladding. No one suggested that firefighters should evacuate buildings and the fire at Knowsley Heights and Garnock Court were not widely known, if at all in London.

I certainly had no idea of the slow dumbing down of standards that was allowing flammable panels to be fitted to buildings. That pace in deteriorating standards was to increase. The temperature for our frog rose dramatically. The first most fire-fighters would learn of any such problems was the Lakanal House fire in 2009.

Part II

From Lakanal to Grenfell

Towards the end of my time at Orpington there was another increase in temperature for our boiling frog. The Regulatory Reform (Fire Safety) Order was due to come into force in 2005. We will cover the details later in this chapter. However, I recall changes before I left Orpington in 2003. Basically, we stopped getting G visits to high-rise blocks. I recall the last one we did and being told by Fire Safety that we were unlikely to get a revisit to check that things had been done. I learned the Brigade was pre-empting the introduction of the legislation by reducing their visits.

Now the onus would be on the owner to inspect and rectify defects in their own buildings. The way it was explained to me was it was intended to be like an MoT for a car. Everyone is responsible for their own car and gets it inspected and tested once a year at a private garage. It would be too onerous for government officials to visit everyone's house and check individual cars.

With high-rise blocks this would free up time for fire safety and mean enforcement could be pursued as and when problems arose. Or, in other words, we would only get involved if it was raised by an owner asking for advice, a third party informing us of a potential fire safety defect, from stations noticing problems after minor incidents (shut in lifts, chutes alight etc.), Community Fire Safety Visits or other ad hoc visits or after a fire.

I don't recall any Fire Safety Officer or any other type of officer being particularly enthusiastic about the new legislation. The general opinion fell into two camps. The first group warned that this would be dangerous. With many owners being unaware, or uninterested in fire safety, standards were predicted to fall. In addition, the introduction of unqualified, ill-defined 'fire risk assessors' was seen as a recipe for disaster.

The second group was a little more sanguine. Their view tended to be that this was a bad thing but that it was possible to make the best of it. In any case, it would indeed free up time and take the pressure and a significant workload off fire safety.

My view on the MoT analogy was that residential high-rises are a little different from cars. Whilst many people do get killed in cars each year, a single high-rise has a much higher life-risk than a single car. Once again, there is nothing to say standards *had* to fall. If everyone followed The Regulatory Reform (Fire Safety) Order correctly, and competent, knowledgeable people conduct regular assessments in a conscientious and professional way, there is no reason why the goals of the legislation could not be achieved.

However, within that is a devil of detail and wishful thinking. Where the analogy with MoTs breaks down is that your car, if taken to a garage, is being serviced, repaired and tested by a car mechanic. Councils could not call on fire safety officers to conduct inspections. Perhaps inevitably, the task of carrying out a risk assessment on a high-rise block was given to someone with little to no knowledge of fire safety.

The result of this change was that standards of maintenance and inspections did indeed drop drastically. The general state of maintenance and fire safety standards in blocks was far lower when I came out of training in 2012 than when I entered in 2003. But the rot had already set in before the introduction of The Regulatory Reform (Fire Safety) Order and it was something I observed and recall very clearly.

In late-2003 I was promoted into training as a Station Officer and posted to Firefighter Development. This meant I would be training new recruits at Southwark Training Centre. I found myself walking under the very same arch I had passed through seventeen years before as a young recruit.

My role as a trainer in Firefighter Development was to deliver the initial training course to new recruits before they went to fire stations. Once at a station, they would start their development period which lasted a couple of years before becoming competent firefighters. The training course at this time was seventeen weeks long and each squad usually had twelve members. It is useful to know what the course did, and did not, contain across the four modules.

Initial Firefighter Development training course

Module one lasted two weeks. In the first week recruits were issued with their uniforms and fire gear. Lectures were given on health and safety and equalities. They were introduced to some basic knots which would be used throughout the course: rolling hitch, clove hitch, figure of eight, chair knot, fisherman's bend, half hitch and others. They might get one chance in that first week to put their fire gear on and familiarise themselves with the equipment and stowage of one of the appliances at the training centre. The second week was a five-day first aid course.

Module two was four weeks of pumps and ladders. This was when the trainees really began to be put through their paces. It was quite intense. By the end of the four weeks, they should all be able to work the pump, tie all the required knots, haul equipment aloft and put the three different types of ladders up safely.

In my experience it took four weeks for a squad of twelve trainees to learn everything, have sufficient time to practise and be assessed. If we use just the example of the 135 ladder, this requires four people to pitch, which means a squad of twelve needs twelve ladder pitches to give everyone just one go in each of the four positions the first time. They then need to do it several times to practise. Eventually, it also has to be done at pace and under pressure and then alongside hose and hauling aloft. In addition to that, there are several types of ladder pitches aside from the basic. A turning drill involves taking the ladder on its side into a narrow space, getting it vertical and then turning it on one corner. Get this wrong and 13.5 metres of 100 kilograms of metal comes crashing down. A 'props to the face' drill involves pitching a ladder over an obstacle and involves taking the props used to steady the ladder round to the front, one by one.

All these also need several pitches to practise and then perform, alongside hauling aloft jets whilst under pressure. Then it has to be assessed and time set aside for any requiring development and re-assessment. This means dozens of drills just with the 135 ladder and a similar number with the smaller 9-metre and short-extension ladders. Learning how the pump works, hauling aloft and getting a jet to work takes a similar amount of time.

Module three, lasting three weeks, was an introduction to breathing apparatus. This was a technical part of the course but also included a number of practical exercises, although none involving fire at this point. It is worth noting how disorientating wearing breathing apparatus can be sometimes. Trainees would often not be able to accurately state where they had been in debriefs. Instructions would be forgotten or misinterpreted when under pressure. One example has stuck with me.

This particular trainee was quite a relatively strong performer. Imagine a corridor with two rooms on the right. His brief was to lead a crew of two and follow the left-hand wall and enter the first room they came to on their left and carry out a search. This meant they would get to the end of the corridor and turn back on themselves and enter the first room on their left (conversely this would be the second room on the right for a crew following the right-hand wall). The area was pitch black and the crew were in breathing apparatus with a mask over the visor to obscure their vision.

They followed the brief and entered the first room they came to. There was a bit of furniture which needed moving and they quickly got to work searching and moving furniture as they went. During the exercise, they finished searching the first room, came back into the corridor, continued on their left-hand wall coming back on themselves towards where they entered the corridor. They ended up searching much of the second room as well. On their way out they retraced their steps with the wall now on their right. They thus passed the first room, turned the corner at the end of the corridor and came back the way they had come in. All very good work except when the team leader was asked for a debrief. It's important to remember that firefighters often do not have a map of the premise and have to draw this outside, based on what exiting crews say and use this drawing to brief oncoming crews.

The crew were adamant that they had only been in one room. They were incredulous when I told them they had actually been in two. So much so that I had to take them down into the basement area and show them. The poor lad was almost in tears of disbelief: 'How could I not know?' he kept saying. He wasn't the first and won't be the last. If it's that easy to be disorientated in a small exercise like that, imagine how easy it is in a complex building with lots of heat and smoke. Then add lots of noise and pressure. And, finally, imagine you're knackered after climbing several flights of stairs and you're low on air.

Back to training and module four contained a host of combined exercises and other skills. Pumps and ladders are combined with breathing apparatus in a variety of exercises. In addition, they had lectures and an exercise in hazardous substances. Some subjects don't lend themselves to a physical drill and so they received lectures on electricity, gas, radiation and special services such as train crashes. In addition to this, there was a week-long road traffic collision (RTC) course where each squad had seven cars which we could use in a variety of exercises: car on all four wheels, car on side and car on roof.

Many subjects are now on computer-based training which means in practice a PowerPoint on an iPad. The RTC course is no longer seven cars or a whole week and some exercises have been scrubbed altogether. Towards the end of the course there was a week of Real Fire Training where they got to do a variety of combined exercises in a real fire situation. This was the first time the trainees experienced real heat and smoke, a vital part of the course. The final week consisted of assessments and preparing for their posting.

A number of topics were considered high risk enough to warrant covering in detail. Importantly, exterior cladding was not among them. One was the modern construction of roofs. Instead of thick rafters and what were known as king and queen posts supporting the structure, modern roofs tended to be made from prefabricated 'triangular pieces'. These were often constructed from relatively thin rafters held together by a strip of metal with a number of metal pins. These pins were 8-10mm long and the plate and pins basically acted as a large staple. The danger from this was that a fire in the roof could burn through a few millimetres of timber and quickly make the entire roof unstable. The metal plates holding the structure together could simply fall out. Hence, we had to show all trainees a video ominously titled, *Seven minutes to collapse*.

The emphasis was that it was quite possible that we could be turning up and gaining entry to a building around seven minutes after the fire started. I am not a building engineer and there may be a very good explanation for these modern methods. But to my untrained eye my old Victorian house with its large rafters, ridge board, supporting posts that I couldn't get both hands round looked like how I imagine a roof should look. Some of the modern constructions, often with no ridge boards and certainly no supporting beams, look like they are made out of matchsticks held together by small metal plates.

This warning video was shown to trainees during the only lecture on building construction they received, which lasted a morning. Ironically, historically the Brigade had a wealth of knowledge to draw from. This was because in the past many firefighters had experience of various aspects of the building industry through their previous employment or second jobs. Most watches would have had an electrician, plumber or builder. Many a job was facilitated by their knowledge from outside the Brigade.

The building construction lecture had a section concerning 'signs and symptoms of collapse'. The lecture was often repeated at stations once a year or so. So, firefighters on average had very little training and knowledge on building construction. Just enough to recognise obvious signs of collapse, such as cracks in walls or lintels, dropped arches over windows or expanding joists pushing out brickwork.

It follows that anyone expecting firefighters to be some sort of quasi-building control inspector is to be disappointed. Nor is it practical to expect them to receive enough training to give them that knowledge, as I will explain later.

However, there is an important aspect connected to Grenfell that I think we certainly should have been aware of. It was only after Grenfell that I became aware of the 1991 Merseyside fire or the 1999 Ayrshire fire in Scotland. The nearest subject I can think of was 'dangers of sandwich panels'. These are often large sheets of metal encasing flammable filler used in many out-of-town commercial units. The filler often burns easily and gives off toxic smoke. In addition, they could lose their integrity in a fire and collapse quickly.

At no stage did trainees or I ever get information relating to the dangers of panels such as were present at Grenfell. Nor was I, or anyone I have spoken to, aware of cladding fires or dangers from exterior fire spread in general. Trainees were certainly told that fire could spread in a high rise. The most common situation was the *Coanda* effect where flames and hot smoke exiting a window would often 'hug' the building and affect the window directly above. Additionally, fire could break out of the original compartment and affect others on the same floor. It was thus possible that people might have to be evacuated from the same floor or floor directly above.

What we taught in relation to high-rise in particular was that fire could break out of the original compartment, but this was rare. In high-rise residential buildings the policy was stay put. We might need to evacuate adjoining flats, but this was also rare. They were taught high-rise procedure and conducted practical exercises in module four of their training. But it focused on the assumption that high-rise procedure worked and implicit in that was the idea that compartmentation was robust.

It wasn't until the Lakanal Fire in 2009 that this bubble was burst. Even then, that fire was viewed with the same incredulousness as the reaction recorded at the start of this book: 'How could this happen?' I can state here it had little to no impact on training in the years between 2009 and the year I left training in 2012. It was almost seen as a one-off freak event. It would be comforting to be able to claim a video existed: 'fifteen minutes until compartmentation across the whole building completely breaks down'; 'fifteen minutes for the fire to jump twenty floors'. Yet no such video or PowerPoint was made prior to Lakanal and even for the years after.

Another topic trainees got no training on was statutory fire safety. firefighters are not Fire Safety Officers. We should first draw a distinction between statutory fire safety and community fire safety. Community fire safety mainly involved home fire safety visits and fitting smoke alarms. Statutory Fire Safety focused on legal obligations. As such Fire Safety Officers have specialist roles in the same way Building Control is a specialist department in the local council. Even at station, the level of statutory fire safety training a firefighter received in London was small. It was usually confined to the Fire Service Regulatory Reform Order or the storage of certain goods such as fireworks or cylinders.

So, in summary, between late-2003 and early-2012, when I left the department, trainee firefighters completed a seventeen-week course where they learned the basics of pumps, ladders, breathing apparatus and a variety of other skills and knowledge. They got no fire safety training and a minimal amount of building construction. They were taught specific dangers, such as cylinders, modern roof construction and sandwich panels, but at no point were the dangers of exterior cladding taught, even after the Lakanal fire in 2009.

After 2012 training was privatised and Babcock took over. The course was soon cut to fourteen weeks, then later to eleven. As I had been the union rep in training, the union asked me to attend meetings on both occasions that management wanted to cut the course. So I had a firsthand look at what, how and why things were done. Suffice to say, my opinion then and now was that this was a regressive step and indicative of dumbing down of standards in general.

Training at station

Having just covered initial training, this is a good point to look at what training firefighters received at station. We mentioned earlier fire safety training and why an in-depth knowledge was deemed not appropriate for station staff. Another reason may be time.

Firefighters work two days, two nights (often called a 'tour') and four days off: forty-eight hours across an eight-day rolling shift. This equates to a forty-two-hour week and forty-five 'tours' a year. That means ninety day duties and ninety night duties. A number are lost through taking leave, a handful more through off-station training, such as mandatory first aid or breathing-apparatus courses. Other slots are used up with large-scale exercises with other stations or agencies.

Within the station training schedule there are some things one has to do more frequently. I would argue that they should have one session doing ladder drills at least once every two months; perhaps knots and lines and pumping as well. It would be useful to have a car to cut up once a quarter (I was lucky to get one once a year at station). You also need to consider that it's often difficult to conduct practical exercises in the evening because of the noise. Thus we tended to confine lectures to the evenings of night shifts and practical exercises in the mornings during day shift.

What this means in practice is that about a quarter of training is taken up by basics such as pumps and ladders, breathing apparatus and RTCs. There are a limited slots for familiarisation with other topics. And, of course, once a year often isn't enough, aside from people missing it due to holidays or sickness. The list of subjects that require training often seems endless: radiation, working of water, hazardous substances, terrorist attacks, IEDs (improvised explosive devices), CBRN (chemical, biohazard, radiological or nuclear attacks), biohazards, train crashes, collapsed buildings, gas incidents, electricity and a host of others.

On top of that, they are also supposed to familiarise themselves with all the different tools. Additionally, they should have training sessions on specific risks on or near their ground. So, if there's a major industrial site, they should make a visit and have a lecture on it. The number of different operational procedures, notes, equipment and local risks greatly exceed the number of training sessions available across a year.

What this means is that most topics are covered by a lecture once or twice a year. The idea that firefighters can have anything but a superficial knowledge of, for example, radiation is wishful thinking. It should be clear from this that firefighters cannot be experts on everything. Core skills, such as pumps and ladders and RTC training, cover the most common incidents. Specialist skills, such as water rescue, line rescue and incidents involving hazardous substance, require specialist crews on fire rescue units. There simply is not enough time to make firefighters at station experts in fire safety, let alone building regulations.

Regarding statutory fire safety, firefighters receive no initial training and when at station might receive a lecture once or twice a year. This will provide a very basic level of knowledge, usually around aspects of The Regulatory Reform (Fire Safety) Order 2005 or storage of hazardous materials, including fireworks.

The lay person may believe the word of an operational firefighter on fire safety matters. Having received an answer, they may go about their business assuming they have an authoritative answer to a problem. This would be a mistake. You may well have the opinion that operational firefighters *should* have a high level of fire safety knowledge. I have no objection to this. The problem is that there simply is not enough time to train firefighters to be experts in every aspect for which the Brigade has responsibility. That's why there are specialist roles. Fire Safety is a specialist role, as is Fire Investigation.

During the inquiry there were a number of questions directed at firefighters concerning building regulations and fire safety. The usual response was to try their best to be helpful. I found myself wishing that, just once, someone would respond a little more bluntly:

> I have no idea. I don't know anything about building regulations and very little about fire safety legislation. I'm not a building control officer or fire safety officer. I've received no training in the former and only have a passing knowledge of some aspects of statutory fire safety. We were always instructed to pass Fire Safety matters on to the Fire Safety Department precisely because operational firefighters do not have specialist knowledge in this area.

But no one said that; they all tried to provide a helpful answer. The media, of course, would portray any lack of knowledge as evidence to cast blame. I've not seen a single article or interview that made clear the difference between operational firefighters and fire safety officers, let alone building control officers.

The dripping tap gets worse

In 2004 the Central Fire Brigades Advisory Council (CFBAC) was scrapped and national standards abolished. This removed central control over policies. Many viewed this as a regressive move. The best thing I can say about it was that at least I was aware of it. Hidden from us at the time was the steady stream of what I can only describe as a series of smoking guns.

To understand the significance of some of the details it is necessary to get a little technical. Bear in mind that, in the UK, building regulations stated that the external walls of the building should adequately resist the spread of fire over the walls and from one building to another having regard to the height, use and position of the building. In the scramble to come up with new methods of construction, this simple fact appears to have been forgotten. Regarding fire classification, the Euroclass system was introduced across the European Union in 2000 and had the following classes:

A1 Non-Combustible
A2 Limited Combustibility
B Combustible materials – Very Limited contribution to fire
C Combustible materials – Limited contribution to fire
D Combustible materials – Medium contribution to fire
E Combustible materials – High contribution to fire
F Combustible materials – Easily flammable

These classes are further divided to provide information on a product's tendency to produce smoke and flaming droplets with three divisions within each:

s1 Emissions absent or very little
s2 Emissions with average volume intensity
s3 Emissions with high volume intensity

d0 No burning droplets
d1 Slow dripping droplets
d2 High/Intense dripping droplets

The year that CFBAC was abolished, Arconic, the firm that supplied the rain-screen cladding panels at Grenfell, conducted a test on its ACM panels in France. The panel came in two types, riveted and cassette. The cassette system failed completely, burning ten times as fast with seven times as much heat and three times as much smoke. It failed to achieve even the lowest grade E.[1] Arconic continued to sell it.

In 2004/5 the now privatised BRE asked government to consider introduction of smoke toxicity limits for materials used in internal walls and ceilings similar to most of Europe. The UK did not follow.[2] By 2005 the EU adopted the new standards from A1 to E. The UK decided to use Euroclass B but retained Class 0 (equivalent to Euroclass C or D) as an alternative.[3]

In 2005 in the UK, Kingspan, the company responsible for much of the insulation used at Grenfell, had its K15 insulation tested at BRE. It was used in combination with heavy cement-fibre cladding panels. BRE stated that the test was for this exact system only. Kingspan marketed it as 'successfully tested to BS8414' and 'acceptable for use above 18 metres'. They did not mention that it was only for one specific system, i.e. with heavy cement-fibre cladding panels. They also marketed it as achieving Class 0 even though it 'was not relevant' and only obtained the pass by ripping off foil normally attached.[4]

Just to emphasise the dangers of cladding, in 2005 there was another cladding fire in a nineteen-story block, The Edge, Salford, Manchester. The panels had a Class 0 rating and had been used in compliance with Approved Document B. Crucially, a subsequent amendment to the passage in guidance was restricted to a paragraph headed 'insulation materials/products', stating 'any filler material' should be of limited combustibility.[5] If this guidance had been clearer and specified that all materials in cladding systems on the exterior of buildings should be non-combustible, Grenfell may have been avoided.

So by 2005 there had been three significant cladding fires. Yet none of this trickled down to either stations or the training department in London. Meanwhile, the anticipated The Regulatory Reform (Fire Safety) Order 2005 came into effect. This introduced ill-defined, unqualified fire risk assessors. I wasn't to see consequences of this until I returned to station years later.

Labour victories in 2001 and 2005 did nothing to stop what we can now see are obvious holes in the system. In 2006, Arconic applied to the British Board of Agrément (BBA) for a certificate for the ACM panels based on the test where panels were riveted. The cassette type test was omitted although the diagram provided showed both types. They also claimed the panels met Class 0 but only provided a 2003 test on a 'fire retardant' version. The British Board of Agrément (BBA) issued the certificate.[6]

This is most important. It meant that a panel, which in one configuration had failed a test catastrophically in France, was able to obtain a certificate implying that it was safe to use on high-rise buildings. The following year a senior Arconic manager attended an industry conference where it was acknowledged that a building covered in ACM would act similarly to a 19,000-litre oil truck and 'kill 60-70 persons'.[7]

Meanwhile, Kingspan were conducting more tests on their own product. In 2007 they changed the chemical formula in K15 and added perforations to the foil. A second test (after the 2005 test) failed despite the use of non-combustible solid aluminium cladding panels in front of the new K15 insulation. It became a 'raging inferno', nearly destroying the lab.[8] Tests a year later also resulted in 'rapid and serious failures' of K15.[9] Despite these test failures, Kingspan, like Arconic, approached BBA for a certificate, but provided only the 2005 test with cement panels. The certificate was issued.[10] The following year Kingspan obtained a certificate from the LABC, London Authority Building Control, which recorded it as 'material of limited combustibility' suitable for buildings over 18 metres.[11]

I can't tell you how appalled I am reading about this years later in Peter Apps' book, *Show me the Bodies*. It reads to me as though standards of testing and certification had completely fallen apart. Instead, we had a hollowed-out system and a box-ticking culture that seemed fine on paper but had little connection with reality. I do not understand the behaviour of either Arconic or Kingspan.

During these years I was at Southwark Training Centre. Course after course was warned of the dangers of cylinders, collapsing roofs and sandwich panels. Course after course practised high-rise procedures. There was no training regarding cladding or evacuation. Many of my former trainees attended Grenfell.

Suffice to say, the seed of a disaster had already been sown. But that did not make it inevitable. Weeds can be pulled up, systems can be overhauled, if only warning signs are heeded. In 2009 we had perhaps the biggest warning sign so far. Unlike the three previous cladding fires this one caused fatalities.

Lakanal House Fire

Lakanal House is a twelve-storey high-rise block in Camberwell, South London. I had been in training for six years when, on 3 July 2009, a fire was caused by an electrical fault in a flat on the ninth floor. As at Grenfell, it suffered from a state of 'low level decay' and 'degradation'.[12] Alterations had been made installing ventilation systems and openings for cables. Some work had resulted in compartmentation being compromised with fire stopping between flats removed. Wedged fire doors and open windows also facilitated fire spread. Window panels had been replaced with 'Trespa' High Pressure Laminate (HPL) panels which burned through in just 4.5 minutes. Debris falling from above enabled the fire to spread downwards whilst heat and flames spread upwards.

The speed, severity and direction of fire took firefighters by surprise. The bridgehead had to be moved further down and, at one point, crews were withdrawn altogether while they attempted to relocate. This left occupants still above the fire. Control continued the advice to say put and were reluctant to appreciate the situation the residents were reporting. Six people died, all on the eleventh floor, including one person who had remained on the phone to control for forty minutes.

Below is an account from a firefighter who attended.

As soon as the bay doors opened and our appliances rolled out we could smell the fire, the Sceaux Gardens estate on which Lakanal sits is directly opposite the fire station, the journey there was quick, right out of the station, left into Southampton Way, left into Dalwood Street and it's on the left. As we approached up Southampton Way and drew level with Lakanal I could see a stream of thick black smoke issuing from the building, this clearly was no cooking pot or barbecue on the balcony.

Due to the number of high-rise buildings on the stations ground this was a scenario that we trained for regularly and a type of incident that we had attended before. Orders were given to go straight into high-rise procedures on arrival, in other words find and connect to a hydrant, charge the dry riser on the building, take control of the firefighters' lift in the building and start loading firefighting equipment into the lift. It was about the same time that a message from control came through on the radio informing me that they had received multiple calls to the fire and were dispatching two additional appliances, E352 Old Kent Road's pump and E381 New Cross's pump ladder.

As we arrived at the access road for Lakanal I could see debris from the building falling in front of us. In order to reach the dry riser the first appliance had to drive through it, debris hitting the roof as it went. The second appliance sensibly parked at the other end of the access road. Both crews deployed from the appliances and we set about implementing the high-rise procedures which all happened in just a few minutes.

The Crew Manager had gone with an entry control officer and breathing-apparatus crew of two to set up a bridgehead. The Pump Ladder was pumping water to the dry riser. At this point in the operations things were from a procedurally perspective going well, the bridgehead had been established, water supplies were secured and a BA team were fighting the fire, It was by all accounts looking as though it was going to be a textbook example of high-rise firefighting, then I looked up. It's not unheard of for fires at high-rise buildings to spread upwards; however fire spreading downwards was a new one for me.

We lay out a firefighting jet and got that to work on tackling the new outbreaks but the jet only just reached the fifth floor and was having a limited effect. I was also trying to get hold of the guys on the bridgehead to let them know that there was also a fire on the same floor as them and below them, obviously not a great position to be in, but radio communications were becoming difficult as there were a lot of people trying to talk on a single radio channel.

Things were starting to get very hectic and busy at the base of the building, additional crews had arrived and the Officer in Charge (OIC) was busy giving them briefs and getting them to work, one of them took over operating the jet from me as we had suffered a burst hose that was supplying the dry riser. The pump operator and I were getting that swapped over. Around this time the first BA crew came back down to the bottom of the building. Two members of my watch, they looked quite happy with themselves, they thought that the fire was basically out as they had been out of loop on developments since they had left the bridgehead. As they had exited the building and walked away with their backs to the building they hadn't yet seen the new fire spread, when I pointed it out to them they were quite obviously shocked.

With the external jet having a limited effect on the fifth-floor fire, and the fire on the seventh floor not being fought at all, it was clear that the bridgehead was now untenable. The bridgehead should ordinarily be situated two floors below the lowest floor involved in the fire, so in this case the seventh floor as the initial fire was on the ninth, we now had fire on the fifth, seventh and the ninth. With the bridgehead located on the seventh, they now had fire on floors above them, below them and on the same floor as them, they had to be moved.

The trouble with moving a bridgehead if BA crews are committed is that the crews are expecting to withdraw back to the bridgehead and find someone waiting there for them; communications with the BA crews were as always difficult; wearing BA is hard work and the facemask distorts voices over and above the distortion caused by heavy breathing from exertion. The OIC at that time made the call to withdraw all crews from the building and regroup on the ground floor and re-start, committing BA crews from there. The opposite side of the building was chosen as there was a large pavement area and no debris was falling on that side. As we were moving round to the other side it became apparent that the fire must have also spread upwards as someone on one of the upper floors was seen on one of the balconies starting to tie sheets together as though they were going to attempt a descent of the outside of the building, obviously a dangerous and desperate undertaking.

There was at this time a Turntable ladder (TL) in attendance and a decision was taken to get it into place on the pavement from where operations were now being conducted to see if it could assist in rescues. However, due to the parking of cars in the street, it couldn't get through; apparently when these buildings were built the vision of private car ownership had not been envisaged. The order to move the cars by force was given and around ten of us proceeded to roll the car out of the way. We literally flipped it on to its roof and out of the way; mission accomplished the TL was able to get into position.

By now the attendance at the incident had increased to about twelve pumps and the officers commanding the incident were calling for as many BA crews as possible, so I dashed back to my appliance and grabbed my BA set before heading back to what was now a large BA holding area at the pavement area. I was put into a crew with two other firefighters and we joined the queue to get a briefing and be committed.

It was whilst we were waiting here that the first casualty was bought out from the building, a woman was being carried out by about four firefighters to waiting ambulance crews. It's not like it is in films and on television, it's almost impossible to put a fully-grown adult across your shoulders when you're wearing BA; the set gets in the way, and it's very difficult to move an unconscious casualty even for a strong fit person. Casualties bend and fold and don't have handles to hold on to; dragging them is the only practical way. I could see that these guys were really struggling to bring the casualty down the narrow stairs.

My crew soon got to the head of the queue and we were given our brief: make our way to the seventh floor, locate a jet from the dry riser and firefight on the north side of the building. We got under air and made our way in and up. On arriving on the seventh floor we located the jet and commenced our task in one of the flats. The flats in this building were of two floors designed in such a way that the top floor sat at 90 degrees to the bottom floor, forming an L shape that interlocked with the flat next door. We knocked down all the flames that remained on the bottom floor of the flat but I could notice that there was far more daylight trying to get through the smoke than I would normally expect. As we got deeper into the flat I could see that the panels that made up the external wall of the flat were totally burned away in places; one trip or slip too close to the edge would have meant a seven-floor drop. We were very careful about where we stood.

The bottom floor dealt with, we then located the internal flat staircase to the top floor where we intended to go next. We were stopped quite quickly in our tracks by a mass of cables that hung from the ceiling right across the stairway. Hanging cables can be a deathtrap for firefighters, very easy to get entangled in them, very difficult to get out of without cable cutters, and back then we didn't carry cable cutters. Having not long before studied a case study with my watch on the tragic death of two firefighters in Hertfordshire, who also died after becoming entangled in cables, this was suddenly in the forefront of my mind – this and our rapidly diminishing air supplies meant that we made the decision to turn around and report back to the bridgehead on the ground floor to be debriefed and pass on the information regarding the hanging cables.

On exiting the building the scene that met us was one of intense activity. It seemed that a large portion of the brigade was now in attendance as well as a good chunk of the Metropolitan Police and the London Ambulance Service (LAS). We were debriefed and then moved off to grab a breather and some water to rehydrate. I remember asking a senior officer if it was worth servicing our BA sets, that is to say change the cylinder for a fresh one so it could be worn again. He replied 'Almost certainly yes.' There was at that point no specific area set up for BA servicing, so myself and a couple of other firefighters went and grabbed a couple of tarpaulin sheets off one of the appliances and a few fresh BA cylinders and proceeded to set one up under a covered walkway so that we could get some shade as well; it was a warm sunny day and we were all pretty hot from our efforts. Some bottled water had also shown up from somewhere, so we grabbed as much as we could and moved that into the newly established BA service area. We then set about preparing our BA sets to get ready to wear again.

It was whilst resting here that I witnessed one of the most tragic things that I have seen in my entire career. A member of the LAS Hazardous Area Response Team (HART) was walking towards the casualty triage area, which was right next to our BA servicing area, performing CPR on a tiny bundle cradled in his arms. It was a three-week-old baby. Sadly, she did not survive. I fortunately didn't have time to dwell on this as they were calling for more BA crews to make their way to the bridgehead which had been re-established on the third floor. Rest over, I made myself available; I was put into a crew and we made our way up to the third floor and were directed into the southern corridor to wait to be called forward for committal.

It was very crowded and we all waited a little nervously, cracking silly jokes to relieve the tension. The gravity of the situation had sunk in and we had no idea what we were going to face when we got up to the fire floors. Eventually, my crew were called forward and we were instructed to make our way to the eleventh floor and search flats off the southern corridor and firefight as necessary. I recall the walk up taking a while and when we arrived on the eleventh floor we took a few seconds to catch our breath before commencing our task. We could see fires behind some doors and above the false ceiling in the corridor; we set about breaking these down with a sledgehammer and Halligan bar that we had bought with us and extinguishing them. We then set about searching the flats; some we had to break into and some had been left open by the fleeing occupants. Fortunately it would appear that all the occupants had made it out as we found no one. The visibility was still quite limited due to the extensive smoke logging. Eventually our air gauges suggested that we should turn around and make our way back to the bridgehead. As we were getting into the stairwell a shout came from behind and one of our crew had become tangled in cables that had fallen from the ceiling and was unable to free himself. Fortunately, the smoke logging in the stairwell was slightly lighter than in the corridors and the increased visibility allowed us to free him without too much effort. We descended the stairs back to the third-floor bridgehead where we were debriefed. We then left the building.

By this time it must have been getting on for about 21:00 hours. It was our turn to be relieved but, as our appliance was blocked in and still supplying the dry riser, the decision was made to walk back to the station carrying our BA sets. The first and only time I've ever walked back from a shout that we had driven to.

In 2013 I attended the inquest. It was not a pleasant experience. I spent many hours in the witness box answering many questions about our response, training, equipment and decisions that were made. I didn't have anything to hide; pretty much everything that didn't go right at that incident was due to the building design and upkeep. We can only deal with what we find in front of us when we attend any incident; we hope that the people responsible for coming up with and enforcing the building regulations are keeping on top of their game. I think everyone involved with the inquiry hoped that enough lessons could be learned from this fire that a similar incident could never occur again. Sadly, the Grenfell fire in 2017 was to prove otherwise.

Coroner's recommendations

Suffice to say we were all in shock after Lakanal. How could we possibly deal with a fire on several floors at once? One where we would have to repeatedly move the bridgehead?

The subsequent inquiry produced a number of recommendations. These included the following:[13] Southwark Council were to give clear guidance concerning the stay put policy especially if conditions changed. They were also to ensure that new residents were aware of the fire safety features, especially escape routes, including the balconies, and fire doors etc. Clear signage and information were also to be provided with provision for non-English speakers. The authority was also asked to consider retro-fitting sprinklers.

Fire Brigades were recommended to improve inspections, noting buildings with unusual layouts and access for aerial appliances. They were also to improve awareness that compartmentation could be compromised, allowing smoke and fire to spread both between compartments and common parts such as stairs and lobbies.

One recommendation was particularly important regarding Grenfell. Approved Document B, a key component of building regulations, was found to be a 'most difficult document to use'. It required constant referrals to other guidance and documents to answer even simple questions. It was thus recommended that it should be reviewed and clarified.[14]

Another key recommendation was to 'consider the retrofitting of sprinkler systems'.[15] A subsequent report found that the cost made them 'not practically or economically viable'. Similar advice regarding sprinklers had been forthcoming after two firefighters died in a high-rise fire in Shirley Towers, Southampton. Yet the government was reluctant. The Department for Communities and Local Government, later headed by Eric Pickles, described fitting sprinklers as 'over-zealous' in adding £13,000 to the cost of new builds.[16] One management body responsible for high-rise blocks that followed this train of thought was the Kensington and Chelsea Tenant Management Organisation responsible for Grenfell.

London Fire Brigade response to Lakanal

The coroner subsequently recognised a number of improvements made by the London Fire Brigade:[17]

- New guidance for crews undertaking risk assessments for sites.
- Guidance on matters to be noted when making familiarisation visits.
- Improved firefighters' awareness regarding fire spread above, below and laterally in a block of flats and procedures for moving a bridgehead.
- Improved communication between control and the incident.
- Improved guidance and training for handling Fire Survival Guidance Calls.
- Introduced a Forward Information Board

Perhaps the most significant addition was a phrase added to policy note 633, *High Rise Firefighting*.[18] The incident commander should 'consider following the evacuation plan devised as part of the occupier's fire risk assessment' and 'it may be necessary to undertake a partial or full evacuation in a residential building where a "Stay put" policy is normally in place'. Yet there was no word as to how this might be achieved. There is no means of communicating with residents unless they call 999. There is no central alarm system. The only practical alternative, knocking on doors, breaks a whole host of procedures.

The quickest way would be simply to send a couple of firefighters to the top floor without breathing apparatus to clear floor by floor with, perhaps, a second crew working upwards to cut the time in half. However, they shouldn't be going above the fire and additionally they should be in breathing apparatus. This would reduce the number of floors a single crew could clear. Given the resources at the start of an incident there simply are not enough people to achieve this. They also should have water with them, and we can only imagine the number of jets and amount of hose clogging the stairs. You wouldn't be able to move for hose.

It read to many of us as a throwaway line inserted to placate an inquiry but with no substance. An empty, hollow phrase designed to cover backs rather than offer any practical guidance. The fact that we never received any practical training regarding the evacuation of high-rise blocks only served to harden that impression. Yet, at the same time, the message that filtered down to stations and training was that this was a one-off freak event, a problem within building regulations that would now be addressed.

In the same year that Lakanal occurred, four years after the introduction of The Regulatory Reform (Fire Safety) Order 2005, the TMO, Tenant Management Organisation, responsible for Grenfell were relying on in-house staff to carry out fire risk assessments. The LFB requested that they appoint a competent risk assessor or receive an enforcement notice.[19] The TMO procured a firm called Salvus who produced a report raising serious concerns. The TMO would later dispense with them and turn to a consultant who might be more willing to 'challenge the LFB on thorny issues'.

A new government

In 2010 the Conservative and Liberal Democrats formed a coalition government. The following year they announced a 'red-tape challenge' which resulted in a resolution to 'kill off health and safety culture for good'. A policy of 'one in, two out' later became 'one in, three out'. This made any new regulations related to Approved Document B and fire safety effectively 'impossible'.[20] We recall that this is the very document the Lakanal Inquiry found needed to be revised.

Over the next seven years the policy of austerity cut fire service personnel by 25 per cent.[21] Across the country in the decade prior to 2021, the number of council inspectors fell by 27.4 per cent.[22] It also resulted in the building control team at the Royal Borough of Kensington and Chelsea (RBKC) falling from twelve inspectors to five.

That same year a fire in a flat on the sixth floor in Grenfell Tower occurred prior to the refurbishment. Compartmentation prevented fire spread. However, the smoke-control system allowed smoke to leak back into the building on the fifteenth-floor lobby. The TMO discovered that vents did not close properly, but delayed repairs until planned refurbishment years later.[23]

In 2011 a further test of ACM panels in cassette form failed again and they were formally categorised as class E making them inappropriate for high-rise buildings across much of Europe. Arconic removed the B grade from their marketing material after a 2012 fire in France but did not mention this change to grade E for cassette type.[24] Similar cladding fires were now occurring elsewhere across the world, such as in the Tamweel Tower, Dubai.

In 2012 the BBA, British Board of Agrément, carried out a review of the 2006 certificate for ACM panels, now reduced from Euroclass B to E. Arconic failed to respond and continued to list only fire-resistant product (which was Class B) on their website. The BBA re-issued the certificate.[25] The following year further tests on ACM panels confirmed the E grade classification of the cassette system and moved the riveted system from B to C. An Arconic employee testified to the inquiry that she did not tell clients in the UK as she did not believe European standards were relevant to the UK.[26]

The end of training

The decision was made to privatise training, a regressive step in my opinion. Babcock took over and, whilst some trainers were seconded for a time, I chose to return to station. There are a couple of points worth making about this transition.

Firstly, management stated that there were no plans to cut the length of the course. As it happened, I had previously shown how the course could be trimmed to fourteen weeks by cutting some superfluous sessions: a bi-weekly 'welfare check' afternoon; a week's worth of gym sessions. I also argued that, having experience of being in a classroom teaching thirty teenagers maths on my own, many of whom didn't want to be there, we didn't need two trainers in a class of twelve adults eager to pass. There were, in short, significant savings that could have been made in the department.

The impression I got was that the organisation was largely incapable of change unless it was politically driven, either by internal or external politics. Additionally, it seemed to me that the drive very early was for privatisation so that any attempt that demonstrated in-house efficiencies wasn't welcome.

Within a couple of years we were told that the course was being shortened. I found myself across the table arguing on behalf of the union with the very people who had previously insisted that an afternoon spent checking on how the trainees were coping was vital. The course was cut, first to fourteen, then eleven weeks. I view this latter reduction as dangerous. Several core skills were reduced. Some topics lost their practical exercises. Other subjects became computer-based training alone, which seemed to be PowerPoint presentations which trainees could ask their trainer about if they had questions.

The other matter regards standards of fitness and strengths. There were two main problems; one involved extending the 135 ladder, the other dragging a casualty. I raise this to highlight the response I got which indicates a mindset that I think is concerning in any organisation, let alone an emergency service.

The 135 ladder is 100 kilograms in weight. It is quite a cumbersome ladder to pitch. The problem arises when it is your role to extend. Normal pitches are not the problem as this involves two people pulling on the line. The props can be rested on the floor and one person holds both props whilst the team leader straddles the bar at the base of the ladder.

The problem arises when the ladder has to be pitched across an obstruction, a car or a ditch (not an uncommon scenario). In this scenario one must do a 'props to the face' exercise. This involves the two prop holders coming round the face of the ladder and holding it vertical with the props off the ground. This leaves one person to straddle the bar at the base of the ladder and only one person to extend it. The problem was that a small percentage of people simply could not extend the ladder on their own. However, the message to trainers was very much that this was a problem management did not want to see.

The Brigade's response was to insist that it was now a five-person drill, despite the fact that pump ladders carrying a 135 ladder often only ride with four people. What finally killed this particular pitch off though was a bit of politicking by the union. An accident in another part of the country cast further doubt on the four-person version. The union official came down and explained that they were going to agree with management that it was now to be a five-person drill because this would force the Brigade to increase rider levels to five on all Pump Ladders.

My position was that this was naive, and the Brigade would simply ditch the four-person drill so as to do away with the need to assess extending a ladder on your own and would never increase ridership on all Pump Ladders across the Brigade. Needless to say, this proved correct. So, from 2012, trainees were no longer taught one of the basic ladder pitches and experienced firefighters were told a pitch they had done for years was now dangerous and that they had to use five people, even if they pull up with four and someone hanging out the window screaming to be saved.

A more serious example of this sort of regressive decision-making concerns casualty handling. The Brigade had long since dumped the firefighter's lift when I entered training. We were not even allowed to teach it, supposedly on health and safety grounds. Instead, casualty handling sessions included the two-person and one-person drag. This was on a live casualty, usually one of the other trainees.

On this note it is worth explaining just how difficult it is dragging an unconscious person. You suddenly realise the meaning of the phrase 'dead weight'. This was the technique for a one-person drag rescue, bearing in mind that the average weights of an adult and an adult male could easily be 15 stones. The rescuer would place the casualty in a sitting position and, from behind, with legs bent and straight back, lock their hands together under the arms and around the waist. Then the rescuer would straighten their legs and lift the casualty slightly off the floor, dragging them backwards so that only heels or lower legs would be on the floor.

Most could achieve this, with few having problems. But then I would get the best performer to try it on me. I would relax my body completely and go as floppy as I could. Normally, those playing casualties cannot but help making it easy for the rescuer, sometimes even sub-consciously. But an actual unconscious person is a completely different matter. The person who had just dragged their 15-stone colleague thirty yards with little problem now struggled even to sit me up. This was a real eye opener for the trainees. Some thought that I was cheating but, in reality, I was simply an inert lump of flesh and bone. They did it on each other again, this time the person playing the casualty completely relaxed and providing no help at all.

Fair to say, most trainees could do it but instead of the rather leisurely drag they did the first time they were now knackered and sweating. This was over a flat clear surface. They quickly realised just how difficult it would be to do this in a flat around furniture. Or up or down stairs. Or if there were multiple casualties and you had to do it twice.

The story I usually told to emphasise the point was of a fatal fire in my early years. I was in a crew of two at a flat fire and we found the fire in a room at the end of the corridor straight ahead of the front door. Perhaps twenty-five feet long. In entering the fire room, we found an unresponsive casualty behind the door as we gained entry. In the thick black smoke visibility was zero and everything was by touch. It was only after we saw that it was a large man, probably over 6 foot tall and perhaps 15 stones. It took an enormous effort to manhandle him from behind the door and into the corridor.

We were already knackered at this point. Also, whilst the fire had been mostly extinguished with a couple of quick bursts of the hose, it was still quite hot. Additionally, the corridor was not wide enough for two people to drag a casualty side by side. In any case I was now in the corridor with my back facing the way out. My partner, Steve Rogers, was still in the bedroom with the casualty between us. I was at the man's head. So I sat him up and managed to lock my hands around his waist, my partner grabbing his feet. Now we tried to half drag, half carry him out.

He didn't budge. We tried again. Adjusted our grip and tried a third time. Barely moved him a foot. Much huffing and puffing and a minute later we had only just about got his legs out of the room. Our Crew Manager, Bill Solis, was at the doorway and shouted down asking if we needed help. All he could hear were muffled noises and curses as we tried to drag him inch by inch. Eventually, Bill took it upon himself to help. We were after all only a few feet away.

So Bill, like many before him, broke procedures and entered smoke without breathing apparatus to assist with a rescue of a saveable life. Between the three of us, we managed to get him another couple of feet. It seemed to take an age. Eventually, I got them to stop. I was at the head so, in theory, I was the one to give the commands in this situation. I adjusted my grip again, but this time stayed on one knee. The other two took a leg each. Instead of trying to drag him in one continuous movement we did the following. I counted 'one, two three' and on 'three' we moved him back just a few inches at a time. 'One, two, three' and another few inches.

In reality, the initial struggle was only a couple of minutes. But in a fire and in breathing apparatus this seemed like much longer. Little by little we got him down the corridor and into fresh air. Worth noting, we were all young, fit and fairly strong. The man wasn't particularly big. Just large and muscular as some men are. Unfortunately, it became obvious once outside that this was a fatal fire as the poor victim showed the effects of the fire although the cause of death was likely smoke inhalation.

Hopefully that demonstrates that the ability to drag a casualty is important. It shouldn't really need to be said. However, as we shall see, it not only has to be said but there was push back against it from within the brigade. The first thing to note is that the brigade simply stopped casualty handling being assessable. In addition, the casualty-handling training session rarely demonstrated or required the trainees to replicate an unconscious casualty. One or two people failing was simply too much for the brigade. Far easier to reduce or remove standards than improve recruitment or accept that some people might fail.

Management's arguments were twofold. Firstly, you should be doing two-person rescues anyway because of manual-handling guidance and health and safety. If you come across two casualties, you should leave one and request a second crew or come back after rescuing the first. Needless to say, the majority of trainers or firefighters don't agree. Of course, there may well be times when it takes two or three to move a casualty as the incident above demonstrated. So the question arises: where to draw the line about what is reasonable to expect?

The Brigade has a 'physical' as part of the entry assessment. Within that is the 'dummy drag'. This is a human-shaped dummy weighing 55kg. This is the weight of an average adult woman. So there is already an acceptance that it is not a requirement to be able to rescue an average-sized adult male on your own. You can agree or disagree but let us leave that to one side. At one point, prospective trainees had to drag this dummy in the method described above.

When the number of people struggling with casualty handling shot up, I looked into what was being done at assessment stage. It turned out that they had started allowing people to drag the dummy by the neck loop used to hang the dummies up on hooks. Not only that, but they moved the test from the concrete yard to a gym with a polished floor. The thirty yards distance seemed to have been forgotten or was ignored.

In short, everything was done to dumb down the test. Some potential recruits who would have previously failed the test were now passing. This simply moved the problem from recruitment into training. Removing any standards in training would thus move the problem into the operational sphere.

As the union rep, I spent a significant amount of time trying to get the test and casualty-handling sessions to be more realistic. I tried equally hard to introduce an element of assessment regarding casualty handling within the seventeen-week course. This all failed. I found out why later.

You might be wondering where all this is going but the punch line is coming. A punch in the guts is exactly how I felt. After all these problems, reports and setbacks, I was invited to a high-level meeting full of heads of departments and senior officers. Just to put it into perspective how out of place I was, my department head, two ranks above me, was one of the most junior there. I was brought along as I'd done various reports on this question and I was the union rep for training.

At the meeting were not just uniformed officers but people from various other departments: equalities, human resources, recruitment, communications team, legal and others. During the meeting the subject of a problem in standards of fitness and strength came up, the main problem being the last mentioned. Lots of talk about ladder extensions, casualty handling and examples of an inability to lift certain items of equipment. Even with the dumbing down of casualty handling and removal of the ladder drill, there were still problems with people not being able to complete certain drills because they could not perform a strength task.

One proposal was simply to avoid having those people in specific positions. But in practice, whether in training or operationally, anyone could find themselves in a position where they might have to drag a casualty or extend a ladder. Or on the third floor having to haul a lot of equipment alone while your colleague handles the jet.

The first response from some was to deny that there was any kind of problem at all. But I had the data. Next came attempts to frame it as evidence that trainers were being deliberately sexist or racist. That was difficult to sustain as there were female trainers and a range of ethnicities. Plus, race wasn't a factor anyway. By the end there could be no denying or dodging the fact that there was a problem with a percentage of people who could not perform various tasks, especially dragging a casualty.

Now came the kicker. Recruitment diversity targets were very important I was told. I countered that so was the ability to rescue a casualty. The argument back was that this was a tiny part of our job and relatively rare. Maybe, I said, but it was an important part and one the public expected us to be able to do. It was one of our most important reasons for being, regardless of how rare it was. Not so apparently. Health and safety was brought up. What about the health and safety of the public?

My main contention was that firefighters should be able enter a premise, locate and drag a casualty out. At the very least, dragging an average adult thirty yards on their own seemed a reasonable standard. Not even downstairs. Just from a rear bedroom of a flat, down the corridor and out into clean air. The counter was that manual handling determined that it was a two-person job. Back I came with occasions when corridors were too narrow. Besides, what if there were two casualties in a situation when a time delay effectively doomed one?

Now the punch line. None of that was important. The Brigade's recruitment targets were more important than the ability to rescue a casualty! The uniformed officers present all took a deep intake of breath. Others squirmed uncomfortably in chairs. I asked her to repeat it. She did. Recruitment targets were more important than the ability to rescue casualties.

So there you have it. This attitude won the day. Now this little anecdote isn't really about recruitment targets per se. It's not even just about dumbing down standards in one area. What it highlights is ideological capture. The type of ideology Gad Saad describes in *The Parasitic Mind: How Infectious Ideas Are Killing Common Sense*.

The public pays taxes and expects a certain level of service from its public services. As organisations evolve and grow and get used as a political football, they attract certain 'drivers' outside their core purpose. People are hired, departments grow. Slowly the organisation turns its eye away from its core purpose and starts prioritising elsewhere. The opinions of the public and workforce are not wanted. Concerns raised are ignored and dismissed. If too vocal and difficult they can be deliberately mislabelled as offensive or inappropriate.

But ask the public and you will find many are not even aware that the London Fire Brigade stopped teaching fireman's lift twenty years ago, effectively banning it. They will be bemused that a one-person casualty drag is not assessed, and appalled at the current 'standards'. But the political class, and a very thin veneer of those in managerial positions receiving significant salaries out of taxpayers' money, are very comfortable telling firefighters and the public that *their* priorities, often politically driven, are more important than the ability to rescue someone from a fire. Or for a crew to pitch a ladder across an obstacle. Or lift the Holmatro cutting tool on your own to make a cut on a car in a road traffic collision.

Now, I am not saying that any of this had a direct impact on Grenfell. Certainly not the casualty handling or pitching of ladders. But I will argue that this shift in attitude does have an affect on an organisation. Importantly, it demonstrates that the Brigade often not only prioritises political fads but will happily dismiss what the public perceives as one of their core functions. I suspect many across the public sector will recognise similar examples.

Back to station

After nine years in training, I was posted to Addington Green Watch. The year was 2012. The Lakanal House Fire had been three years previously. The refurbishment of Grenfell was a little more than a twinkle in the eye of the local authority. The Regulatory Reform (Fire Safety) Order 2005 was seven years old.

The Regulatory Reform (Fire Safety) Order 2005

We have noted that, prior to 2003, the Brigade was already preparing for the introduction of The Regulatory Reform (Fire Safety) Order 2005 by reducing visits from station staff. Two years after I came into training, the Order came into effect and replaced the previous Fire Precautions Act (1971). It is worth looking at this in more detail as it is relevant to Grenfell Tower.

The Fire Precautions Act was itself the result of the evolution of fire safety legislation. Prosser and Taylor provide a list of this legislation prior to 2005.[27] Many of these acts were a reaction to major fires and tragedies. As such, fire safety can be said to have been reacting to events, rather than being a proactive approach, looking forward.

The Fire Precautions Act (1971) was not without complaint. A complex, lengthy process and costly to business on one hand and difficult to enforce, with limited prosecutions, on the other. With the evolution of the above pieces of legislation, it is understandable that government wished to simplify

Table 1: Fire safety legislation prior to 2005.

Legislation	Comments
The Factories Act (1937)	Required a means of escape in a building with more than twenty people employed, or where ten people were employed over twenty feet above ground, or where explosives or highly flammable material were stored.
The Fire Services Act (1947)	Gave fire services responsibility for providing advice and assistance on fire prevention matters, the local authorities retaining the responsibility for enforcement.
Factories Act (1959 and 1961)	Issuance and enforcement of escape certificates became the responsibility of the fire authority.
Licensing Act (1961 and 1963)	Brigades required to inspect clubs, ensure safety measures were in place and that there were adequate fire exits and extinguishers.
The Offices, Shops and Railway Premises Act (1963)	Requirement for means of escape, certification by local authority, maintenance of means of escape, provisions of fire alarms, extinguisher and training to staff and powers to magistrates to prevent use.
Fire Precautions Act (1971)	Requirement for designated premises to have a fire certificate. Single private dwellings and places of public worship exempt. Initially applied to hotels and boarding houses with accommodation for six or more people or accommodation above first or below ground floors.
Fire Precautions (Factories, Offices, Shops and Railway Premises) Order 1976	Fire certificate required for premises with twenty persons or more employed or ten persons working on other than ground floor or where explosives or highly flammable materials stored or used.
Fire Certificates (Special Premises) Regulations 1976	Responsibility for issuance and enforcement of fire certificates for major chemical plants and other high-risk premises assigned to HSE
Fire Safety and Safety of Places of Sports Act (1987)	Requirement for safety certificate from local authority. Guidance relating to entrances, exits, stands terraces, ramps, stairways, crush barriers, handrails, perimeter salsa and fences introduced.
The Fire Precautions (Workplace) Regulations 1997	Required fire safety risk assessments alongside previous requirements.

things. The legislation listed above evolved over a number of decades, creating a complex web of overlapping legislation. The Fire Safety Regulatory Reform (Fire Safety) Order 2005 was different in that it wasn't prompted by a disaster. Rather it was the result of pressure from various stakeholders, including unions, fire services and businesses.[28]

The new legislation did a number of things. Firstly, it took the burden of fire certification off local authorities and fire authorities. Its effect was to put the onus on building owners. The order applied to all premises, apart from single private dwellings. It required an 'identified responsible person' to carry out a fire risk assessment. This had to be written down if there were more than five people employed. In theory, this meant that a small shop with one or two people had to have a fire risk assessment but did not need to write it down.

The Fire Risk Assessment was the cornerstone of the order. In essence it followed some simple steps:[29]

- Identify fire hazards.
- Identify people at risk.
- Evaluate, remove or reduce risk and protect against remaining risk.
- Record findings and actions.
- Review and revise.

In addition, the 'responsible person' had a number of responsibilities to provide the following:[30]

- Measures to reduce risk of fire and fire spread.
- Measures in relation to means of escape.
- Measures in relation to means of fighting fires.
- Measures in relation to means of detecting fires and giving warning.
- Measures relating to instruction and training of staff.
- Measures relating to mitigating the effects of fire.

In theory, the Fire Safety Regulatory Reform (Fire Safety) Order 2005 is enforced by the local Fire and Rescue Authority. The first step would be a discussion with the 'responsible person'. It is only if that person has not been 'broadly compliant' that the following action is taken:[31]

- Improvement letter.
- Action plan.
- Enforcement notice.
- Prohibition notice.
- Prosecution.

However, in the following years there was a reduction in enforcement staff:[32]

- Reduction of 40 per cent of fire safety audits in England within 10 years.
- Loss of 17 per cent of firefighters between 2014-2019.
- Reduction of fire safety staff by 35 per cent.

This has led to the criticism that a 'light touch' fire safety regime has led to a 'fire risk assessment gold rush' and a 'consultants charter'.[33]

My own experience comes in three parts. Prior to coming into training, I was a watch manager when preparations were put in place for its introduction. The general opinion was that it would herald a deterioration in standards and, in my experience, that fear appears to have been proved correct. We can see a good example of this with the residents' complaints about Grenfell. The point I would make is that their comments are not untypical.

Around 2010 I was seconded to Fire Safety for a couple of months as training had a period of fewer courses than normal. Whilst there I accompanied several different Fire Safety Officers in visits to residential high-rise blocks across South London. I was able to see for myself the standard of maintenance and compare it to when I conducted similar visits as an operation Watch Officer. In general, I would describe the standard as very poor and much worse than in 2003. When I asked why that was, I received a consistent message.

The local authorities had little money and, to a lesser extent, less experience, knowledge and inclination to maintain a high level of inspections or fire safety standards. The Brigade, as the enforcement agency, would request improvements. The Local Authority would plead poverty. The Brigade would try to force the matter. Somewhere in that process the Brigade would be asked, or forced, to back down because there was no funding for repairs and rehousing entire blocks was impractical. Thus, a situation was allowed to develop whereby, so long as a defect was on a 'to do' list, the situation was accepted, even if that defect took months to fix. Some apparently remained on the list for years.

One reason may be that the cost of litigation makes Fire and Rescue Authorities reluctant to take too many cases.[34] My own experience talking to Fire Safety Officers backs this up. They might have a number of cases on their desks but only enough time, money and resources to pursue one. And that one takes a lot of time going through the court system. The result of this is that only the most serious cases make it to court. Usually after an incident.

An experienced Fire Safety Officer, since retired, told me the following disturbing practice. Fire Safety Officers at one office were encouraged to conduct fire safety audits sitting in the office via google earth. This allowed them to assess a relatively large number of low risk premises quickly and thus 'tick off' more jobs. Of course this in no way replicates an actual physical on-site visit. But the pressure was to complete a certain amount of work with less resources in an environment of reduced enforcement powers.

I suspect the focus on quantity and not quality was largely because one can measure the former more easily. An officer conducting multiple visits to one high risk site and attempting to go through hoops of serving notices might well get admonished for not achieving a target. Yet he or she might have done more meaningful actual fire safety work than another officer, as he put it, 'google-earthing shops'. However, there seemed to be little political will within the organisation to put quality above targets. One wonders what might have happened in the 2000s if the London Fire Brigade had served notices on several Boroughs forcing them to rehouse a high-rise block or two due to poor fire safety. No doubt this was difficult to do politically.

Another issue was the push to make buildings more environmentally friendly. Adding insulation to buildings and covering them in panels seemed to be positively encouraged. Good quality building regulations, fire safety legislation and enforcement standards would ensure everyone involved were well aware of the importance of compartmentation and the importance of not placing flammable material over the exterior of a block of flats. As we saw in the inquiry few, if anybody, of those involved in the refurbishment had the faintest idea of, or cared about, these issues. That's bad enough. But functioning building control and fire safety should pick this up. The fact that it did not shows that we have allowed these services to became less than effective to put it mildly. The fact hundreds of similar panels were passed by scores of building inspectors on hundreds of other buildings over decades demonstrates the problem is systemic.

The final part of my experience comes from a series of audits I conducted on local shops and businesses, including an industrial estate, on my new station's ground. I visited several dozen premises. Of these, around a third had never heard of the legislation: not just didn't know its contents or their responsibilities but were not even aware it existed. That was around 2016-

2017. Of those who were aware, very few had a good understanding and could show me a fire risk assessment. So there we have a key piece of fire safety legislation and the general understanding and compliance appeared extremely poor. If we return to our MoT analogy this would be like a third of car drivers being totally unaware of what an MoT is, let alone the requirement to have one.

Lakanal House fire inquest

In 2013 the inquest into deaths at Lakanal House fire recommended Approved Document B be reviewed and gave clear guidance, in particular to the spread of fire over the external envelope of the building. It also advised government to encourage consideration of the retrofitting of sprinklers in high-rise residential buildings. The government restricted changes to a certification scheme for window installers (ignoring the recommendation concerning the entire external envelope of the building). Regarding sprinklers, they simply reiterated advice following the Shirley Towers fire in Southampton. The reason given for not pushing sprinklers was a concern of being legally required to provide funding.[35]

From my own point of view a Lakanal-type scenario was the one incident that concerned me most. I remember sitting my new watch down and going over Lakanal very soon after I got to Addington. The Brigade would take another year to produce a training package, so we had to do our own version. We even did a small-scale exercise at a local block to see if we could get crews to various flats in a reasonable time using the fire survival guidance calls procedure.

I can't say I was particularly impressed with the new forward information board. We felt it would be preferable to have something far simpler. Or even a couple of 6-feet-by-3-feet whiteboards on which we could quickly draw a building plan and tick off where people were trapped, and what flats or floors had been searched. There's a reason firefighters resorted to writing on the walls at Lakanal and Grenfell. A large blank space is ideal.

Of course, Lakanal was the only cladding fire we were aware of and we certainly didn't envision a Grenfell-type incident. Our conclusion was that if we ever had a situation where we needed to evacuate a high rise, there was no way we could do it keeping to procedures. The next question follows: how to achieve it by bending or breaking the least number of procedures?

The only answer we could come up with to start evacuation quickly was this: a crew of two, perhaps wearing breathing apparatus but not started up, would quickly go to the top floor and alert the residents and then come down floor by floor. If the first crew was already committed into the fire, then you might only have one spare person, not in breathing apparatus, who could run upstairs and bang on doors.

Of course, this assumes that the often only staircase remains free of smoke, let alone heat. One story I heard involved a high-rise incident in south London after Lakanal. A block had become smoke logged. The Watch Manager decided the block needed evacuating. The crew manager, without breathing apparatus, led a crew of two, with breathing apparatus but not under air, up the stairs. The crew had their masks around their necks and first breath lever depressed temporarily cutting off their air. At each floor the crew were sent down the corridor banging on doors. They only donned their sets and went under air if needed. The crew manager informed the Watch Manager as each floor was cleared. This was achieved quickly.

It would have been difficult to achieve this with one crew on one cylinder if under air the whole time. Impossible to do it speedily with multiple crews. Multiple crews were not available. Multiple crews are never going to be available in the first five minutes of an incident. It is unthinkable to achieve this evacuation if each crew needed to set up a line of hose. Besides, this was a one-appliance station so there was an initial crew of five. Neither management nor the union would view this as good practice. However, I am not sure what the alternative is in the early stages of a fire with limited resources available?

As time went on, Lakanal faded into the background. I began to think that perhaps it really was a one-off. Hopefully, whatever allowed the panels to be fitted at Lakanal had been sorted. Meanwhile our dripping tap continued to send standards swirling around the drain.

Drip, drip, drip

The year 2014 was rather busy for our sorry tale. The government was advised by an All-Party Parliamentary Group to take into account new research, abandon Class 0 and, also, that sprinklers were now more cost-effective. These could all be dealt with by 'simple amendments' before a full review scheduled for 2016/17. The government refused. Dr Webb, expert advisor to committee, wrote warning of 'a major fire tragedy with loss of life'.[36]

That same year a Celotex insulation product, later used on Grenfell, failed a test at BRE with cement-fibre cladding panel. In a second test, temperature monitors were reinforced with fire-resistant boards. This passed and allowed Celotex to claim the product was suitable for buildings over 18 metres. They obtained a certificate from LABC.[37]

Kingspan secured a test pass but it was on a different material. Kingspan sought advice from Dr Barbara Lane (later to be expert witness to inquiry). She told the National House Building Council (NHBC) that she was 'deeply concerned … the use of highly combustible materials in residential buildings is now simply an accident waiting to happen'.[38] A senior civil servant e-mailed a senior figure at the NHBC, warning of the potential dangers of plastic insulation used in cladding panels.[39] This did not prevent the NHBC from stating that if no test information was available a desktop study report from an expert would suffice.[40]

Meanwhile, the refurbishment at Grenfell was underway. The project manager at the TMO for Grenfell recalled the Lakanal House fire and requested asking for clarification regarding 'flame retardant requirement'. There is no evidence of a response or follow up and this was described in the inquiry as the 'last chance to avert disaster'.[41] The London Fire Brigade issued a deficiency notice in March, ordering the TMO to fix the smoke-control system by May. Work was delayed for another year.[42] By March, the TMO had a backlog of 1,400 incomplete actions and 650 blocks requiring risk assessments.[43]

Meanwhile, the London Mayor, Boris Johnson, cut ten fire stations, twenty-seven fire engines and 552 firefighters. Before we move on, I would like to make a point about this. Politicians cite the reduction in fires as a reason to cut resources. Firstly, the reduction in fires is often partly down to fire prevention, community fire safety and statutory fire safety work done by brigades up and down the country.

Cutting resources after the success of public services is a bit shortsighted. In addition, it takes just a few minutes to die in smoke and one must ask what a fire service is for. If it is to rescue people in the event of fire, then you need fire stations close enough to attend in a reasonable time. You could have zero fires and close every station in London, leaving just one in the centre.

But that wouldn't be a rescue service in the event of a fire. That would be a body retrieval service.

Additionally, politicians seem to forget what the fire and rescue service does. Like most people they just think 'fire' and nothing else. But the fact is that the Fire and Rescue Service rescues more people from road traffic collisions than fires. Many times more. In that context we are taught that it's vital to get oxygen onto a casualty within fifteen minutes of the incident. It's also vital to get the casualty to hospital within an hour. Again timed from the impact.

Figure 3: Grenfell Tower before refurbishment. (*Wikimedia Commons*)

Thus the Fire and Rescue Service has to have stations positioned so that they can attend in a timely manner. If the ambulance takes twenty to thirty minutes to get the patient to hospital, this leaves thirty to forty minutes for the brigade to receive the call, attend and extricate the patient. Extrication can take from a few minutes to much longer. You certainly can't achieve this within an hour if too many stations are closed.

Lastly there's all the other services politicians seem to forget Fire and Rescue Services provide: train crashes, aeroplane crashes, chemical incidents, radiation incidents, biohazards, terrorist attacks, gas explosions, collapsed structures, tunnel rescues, water rescues, flooding and a whole host of other incidents the other emergency services have not the resources, training or equipment to deal with.

By 2014 most, if not all, station-based firefighters were still unaware of just how bad the situation was regarding cladding panels. The management response to Lakanal was to tell us to 'consider evacuation' but give no guidance or training as to just how that was to be achieved. The steady deterioration of standards across a range of related areas was coming together in a perfect storm brewing in west London.

Grenfell Tower

Building approval for Grenfell was granted in 1971. One staircase was deemed sufficient. Fire alarms and emergency lighting were not required as simultaneous evacuation of the entire building was not required.

Buildings greater than 18 metres in height should have a 'firefighting shaft' containing a firefighting lift, firefighting stairs and a firefighting lobby. This provides a protected area from which firefighting operations can take place. This shaft is normally protected by sixty minutes of fire resistance. The lobby and stairs of the shaft should have a method of removing smoke and heat. This is a vital component of the fire safety of the block. Any residents following the fire safety guidance of the block would have to pass through the lobbies and main staircase.

Being over 60 metres in height (and 50 metres after 2006), Grenfell should have had a wet rather than a dry riser. A wet riser is attached to a water tank, usually on top of the building, and is permanently 'charged' with water. This is because it becomes impossible to provide sufficient pressure above a certain height by pumping water into the dry riser inlet at ground floor level via a fire engine. One would have expected this oversight to be corrected during the refurbishment of 2012-2016. However, the dry riser remained.

The block itself was completed in 1974 and was typical of many high-rise blocks in the country. It originally contained 120 flats spread over 24 floors and a basement. Each floor of the upper twenty levels contained six flats: four two-bedroom flats and two one-bedroom flats. This was built around a central core which contained two lifts, a central staircase and a rubbish chute.

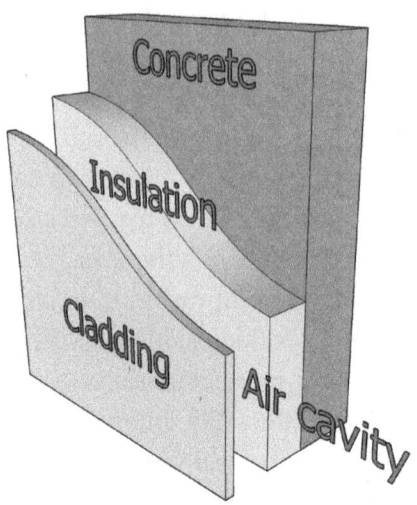

Figure 4: Cladding and insulation at Grenfell. (*Wikimedia Commons*)

Refurbishment began in 2013 and was completed in 2016. The commissioning agent was Kensington and Chelsea Management Organisation (KCTMO). Funded by the Royal Borough of Kensington and Chelsea (RBKCC) which also acted as the Building Control Authority (BCA). The decision to change the non-combustible material within the cladding with a polyethylene core was approved in September 2014.

UK building regulations state that the external walls of the building should adequately resist the spread of fire over the walls and from one building to another, having regard to the height, use and position of the building. The document *Fire Safety in Purpose-Built Blocks of Flats*, Local Government Association Group 201, reiterates this specifically in relation to cladding: 'external cladding should not provide potential for extensive fire spread'. Yet the inquiry Phase one report was to find: 'The external walls of the building failed to comply with ... the building regulations 2010, in that they did not adequately resist the spread of fire .... On the contrary they actively promoted it'.

Let us first look at this cladding: the 'ventilated rain-screen system' sheltered the building from rainfall and left gaps for ventilation and drainage. It consisted of, first, one or two layers of Celotex polyisocyanurate (PIR) foam insulation or Kingspan phenolic polymer foam. These were attached to the original concrete wall. Both have comparatively low times to ignition and support rapid flame spread. They also insulate the cavity from loss of heat.

A gap separated this inner layer with the outer screen. This latter facing layer consisted of aluminium composite panels (ACM). These panels were effectively a sandwich, a polyethylenene (PE) core sandwiched between two thin aluminium panels.

It was this PE core that was the major contributor to the fire spread. It melts, drips and burns. According to one expert, Luke Bisby, 'any competent fire professional could have been expected to be aware of these dangers'.[44]

There were other problems regarding the panels aside from the fact that they were a solidified hydrocarbon and behaved accordingly. The ACM panels, manufactured by Arconic, were cut and folded into 'cassettes' to be hung on rails. Any exposed edges would allow access to the flammable core. The rails allowed pooling of melted PE. The cassette-type system was the very system that had failed a test in 2004, burning ten times as fast with seven times as much heat, three times as much smoke and failing to achieve even the lowest grade E classification.

Sealing the exposed edges of these panels might have reduced fire spread. But perhaps more significant was the absence of adequate vertical and horizontal cavity barriers. Inspections after the fire found they were not continuous and were often poorly fitted, leaving gaps between them.

But, of course, cladding was not the only problem. The windows had been removed and replaced with smaller ones fitted flush to the new cladding system. This effectively moved them outwards by several inches, leaving a gap between the old concrete building face and the new position of the windows, flush with the outer layer of cladding. Cavity barriers should have been placed around the windows to prevent fire spread. Unfortunately, none were present. Worse than that, in places the gap was filled with extruded polystyrene, flammable phenolate foam and spray foam, and ethylene propylene diene monomer rubber. Dr Lane, expert witness to the inquiry, stated that once a fire inside a flat reached a window 'there was a real likelihood' that fire would spread from the flat to the cladding.[45]

This affected the fire in two ways. Firstly, it contributed to the initial fire spread from the kitchen fire to the cladding system. Secondly, it allowed a pathway from the exterior cladding fire back inside individual flats. Initially this was vertical and affected flats directly above the initial fire compartment in flat 16. The architectural crown was also wrapped in ACM panels with exposed edges of PE found all over.

The building was covered in flammable insulation which in turn was covered by panels containing basically solidified fuel. In between these two flammable coats was a gap which acted like a chimney. Aided by the absence of adequate cavity barriers, a helpful pathway was provided by the creation of a gap when the windows were pushed outwards, and that gap was filled with flammable materials. This same pathway helped the fire re-enter other flats, first vertically, then horizontally.

All that is bad enough but there was more. The front doors, which should all be fire doors, had been replaced between 2011 and 2013. As we have seen, fire doors are one of the most important features of the fire protection of a building. They limit the spread of fire for a specified length of time and protect the escape routes to the final exit. This protects both occupants and firefighters gaining access and fighting the fire. The fire door is arguably 'the most important fire safety measure in a block of flats'.[46]

It was found that the majority did not close properly and did not resist fire for the required thirty minutes, some lasting only fifteen minutes. This allowed fire and smoke to spread from a flat into the lobby. Now we have a path from the original kitchen fire to the panel system, up the building, back into a different flat and into the lobbies on several different floors in a short period of time.

Ironically, the block was much safer before they had done any of this work. A fire in a flat a few years before was extinguished without incident. The refurbishment was completed in 2016. It is very difficult to see how any significant fire in any of the flats would not have resulted in flames gaining access to the PE core, Equally, it's difficult to imagine how such a fire would not develop as it did at Grenfell.

Table 2 lists key moments on the timeline.[47]

Table 2: Timeline of refurbishment at Grenfell.

Date	Comment
1 Nov 2011	RBKC and KCTMO discuss plans to refurbish tower.
29 Nov 2011	Architect Studio E asked to carry out work despite no experience on high-rise.
4 Apr 2012	Studio E selects solid zinc panel for cladding. At meeting with CEP Architectural Facades cheaper ACM panels are suggested.
16 Aug 2012	Consultant e-mails Studio E regarding insulation made of combustible plastic. Neither are aware it did not meet building guidance standards.

Date	Comment
31 Oct 2012	Fire safety consultant Exova in fire strategy for the refurbishment plans makes no mention of plans to clad the building and says the work is expected to have 'no adverse effect on the building in relation to external fire spread, but this will be confirmed by an analysis in a future issue of this report'.
26 Feb 2013	Project appeared £2m higher than previous estimate. Studio E suggests ACM panels as option.
April 2013	Report reiterates project is over budget and could fail.
21 May 2013	Meeting between KCTMO and RBKC decides value for money to be key driver for project.
27 Sep 2013	Studio E meets cladding sub-contractor Harley Facades who recommend ACM panels
5 Nov 2013	Exova produce final version of fire strategy, containing no reference to cladding but says design will have 'no adverse effect' in relation to external fire spread but that this will be confirmed in future analysis.
21 Nov 2013	Studio E prepare specs for job. Celotex is listed as insulation material. Solid zinc and aluminium panel listed for cladding but with Reynobond ACM as one of 3 alternatives.
6 March 2014	3 contractors submit tender bids. Rydon at £9.249 million are lowest.
18 March 2014	Ahead of formal award KCTMO meets with Rydon, despite legal advice that conversations of this nature were prohibited before the contract was awarded, to discuss reducing costs to £8.5 million.
31 March 2014	Rydon is successful contractor to design and build refurbishment. Harley is cladding sub-contractor. Study E retained.
8 May 2014	Design team request permission from RBKC planning department to change from zinc to ACM panels. Certificate from BBA suggests panels have Class 0 classification is seen. Planners approve switch and insist on cassette-type system.
June 2014	Mock up of cladding system installed but uses panels with fire retardant core. Eventual cladding had far more combustible polyethylene.
4 Aug 2014	Application to building control for 'full plans approval' which would confirm planned work complies with building regulations. Drawings sent on 24 Sep show cladding wrongly still marked as zinc.
18 Sep 2014	Architect at Studio E e-mails Exova with drawings from Harley asking for guidance on correct positioning on cavity barriers. Advice that 'if the insulation… is combustible you will need to provide cavity barrier as shown on your drawing' is wrong as combustible insulation should not be used.
12 Nov 2014	Project manager at KCTMO recalling Lakanal fire e-mails Artelia and Rydon querying cladding regarding flame retardancy. No record of reply.
18 Nov 2014	Building control officer e-mails Studio E giving full plans approval 'subject to conditions'. No record of these conditions have been found.

Date	Comment
17 Mar 2015	Harley Facades purchases Celotex insulation from supplier SIG.
27 Mar 2015	In a debate about fire stopping, a designer at Harley Facades states: 'There is no point in "fire stopping", as we all know; the ACM will be gone rather quickly in a fire!'
30 Mar 2015	Siderise, sub-contractor of cavity barriers e-mails Harley Facades design drawings highlighting a 'weak link for fire' at the top of the windows. The e-mail is not forwarded.
6 May 2015	E-mails show combustible Celotex insulation has been selected to fill gaps around windows, despite non-combustible Rockwool having been specified.
26 May 2015	Harley Facades orders Kingspan K15 insulation to supplement low availability of Celotex. KCTMO are not informed of switch.
22 June 2015	Rydon writes concerning 'poorly performing site'. Later inspections fail to spot flaws including badly fitted or missing cavity barriers.
29 July 2015	Project delayed and behind schedule for target October finish date.
9 Oct 2015	Legally required handover documents from Artelia to KTCMO 'only been partially completed and is never fully done'.
9 May 2016	With project overrunning consultant at Artelia writes to Rydon stating, 'This is becoming a farce ... I do not think I have ever worked with a contractor operating with this level of nonchalance.'
13 May 2016	KTCMO press release celebrates completion of refurbishment.
7 May 2016	RBKC building control issues completion certificate for project.

Dumbing down of testing standards is one thing. Enforcement of testing standards is another. As is monitoring, inspection and certification. What we see above is a whole other layer. Here is a whole host of people who should be aware of the importance of compartmentation. Yet I get the impression that a significant number of people involved in this project had no idea that the entire fire safety of the building rested on compartmentation holding. I would have expected anyone who did know better to be mentioning it in every other e-mail: 'Don't forget cavity barriers, don't forget any gaps around windows have to be sealed', etc. Instead, the evidence from the inquiry showed that too many people had zero understanding of this vitally important concept.

And so it continued...

Meanwhile, in 2015 the NHBC had written to Kingspan warning K15 insulation was to be rejected for use on high-rise blocks. Kingspan responded with a solicitor's letter warning of legal action. The NHBC moderated its position and allowed for desktop study. The following year the NHBC published guidance setting out systems that were acceptable, even without a desktop study. This included Kingspan's K15 and Celotex's RS5000 and even ACM panels, so long as they had a Class B rating.[48]

That same year, after a fire at nearby Adair Tower, the London Fire Brigade issued the TMO in Kensington and Chelsea with a 'deficiency notice' demanding that all doors have self-closers. This was followed by an enforcement notice.[49]

In 2016, the year the refurbishment at Grenfell was completed, a civil servant in the department responsible for building regulations admitted in writing that ACM panels were flammable when the polyethylene core is exposed.[50] A cladding supplier also wrote to the government department with 'grave concern' regarding ACM, combustible foam thermal insulation boards and the number of buildings involved in the UK. The response insisted that the rules regarding flammable material on the exterior of high-rise buildings are not ambiguous and 'if designer and building control body choose to use them it is up to them'.[51]

A group representing the cladding industry agreed Approved Document B required clarifying regarding materials used on the exterior of buildings. An all-Party Parliamentary Group on Fire Safety continued to write to government with concerns. A Dr Webb raised the matter with a senior civil servant responsible, warning of a death toll ten to twelve times that of Lakanal (i.e. sixty to seventy-two people). Despite all this, a government report concluded there was no need for change.[52]

Arconic warned its French team of significant differences between PE and FR panels and required FR to be used in future. A similar warning was not given to the UK.[53] It is the PE panels in cassette form that were fitted to the exterior of Grenfell Tower.

Back in the UK the London Fire Brigade carried out an inspection of Grenfell Tower and issued a deficiency notice regarding fire doors not self-closing. Investigations after the fire found 43 of 129 doors had no self-closers and 34 that did failed to work properly.[54] A report in 2022 found that 75 per cent of 100,000 fire doors were defective in some way. At Grenfell the doors failed in as little as seven minutes.[55]

In the April of 2017 the Brigade Commissioner, Dany Cotton, wrote to government ministers warning of blocks with 'significant compartmental issues'. The RBKC sent the letter from the LFB to the TMO who forwarded it to a consultant risk assessor. They believed that they did not have such cladding. He replied that the cladding complied with building regulations.[56]

At station level we were all still unaware of such deep-seated problems with building regulations. By 2015 the Lakanal fire was six years in the past. If someone had asked me at that moment what the situation was regarding residential high-rise, I might have commented on a general deterioration of standards of maintenance and cited the introduction of The Regulatory Reform (Fire Safety) Order 2005 as a pivotal moment. I was not aware of the more serious problem of compartmentation. The question of flammable panels at Lakanal did indeed seem to be a one-off.

Even when we saw a cladding fire at a tower block in Dubai, few of us thought a Grenfell-type incident might happen here. A fire broke out in the building at 2:00 am on Saturday 21 February. Hundreds of people were evacuated from one of the world's tallest residential buildings when a blaze swept through the seventy-nine-storey skyscraper. External cladding was charred from the fiftieth floor to the top of the tower. I recall discussing it at station and the general consensus being that was 'their Lakanal'. A combination of a feeling we had got off quite lightly in 2009 and thankfulness that such an incident was unlikely here.

I was rather shocked when I discovered at the inquiry that a PowerPoint presentation had been seen by senior officers detailing a number of cladding fires at home and abroad. Before we turn to that, there occurred an incident that completely changed my view.

Shepherd's Court fire

On 19 August 2016, twenty fire engines and an aerial ladder platform attended a fire that originated in a two-bedroom flat located on the seventh floor of a purpose-built block of residential flats of twenty floors in Shepherd's Bush. A mother and son were at home when a fire started in the tumble dryer. When firefighters arrived, they were confronted with a flat fire that was rapidly spreading up the outside of the building from the seventh floor. Fire engines were increased quickly to four, eight, twelve, before finally being increased to twenty. The fire spread vertically up the exterior cladding, involving several additional flats via open windows. The insulation within the panels was polystyrene.

The idea that the problems that caused the Lakanal fire had been dealt with was now dispelled. I recall talking with the watch about it. There seemed a real prospect of another Lakanal. We mentally dusted off our plans to bend a few procedures. In reality, we may have been half-prepared for another Lakanal but we were ill-prepared for anything like Grenfell.

Tall Building Facades PowerPoint

As we pondered the ramifications of the Shepherd's Court fire at station, unknown to us a PowerPoint had been seen by some senior managers. A London Fire Brigade PowerPoint presentation dated 13 July 2016 entitled 'Tall Building Facades' featured a number of cladding fires, including the Granock Court fire in 1999. Fires abroad included Grozny, two in Dubai, and one in Baku, Azerbaijan in December 2015. It is noted that flames in cavities can extend five to ten times their original length, regardless of the materials present. Much of the information appeared in a similar presentation by Stephen Howard, Director of Fire testing and Certification, BRE dated to June 2016. Titled 'BRE: The Fire Performance of Building Envelopes', it listed a number of fires involving 'fire spread in building envelopes'.[57]

- Knowsley Heights 1991
- Basingstoke 1992
- Irvine 1999
- Paddington, London 2003
- The Edge, Manchester 2004
- Windsor Tower
- Madrid 2005
- Berlin 2005
- Hungary 2009
- Dijon France 2010
- Chechnya
- UAE
- USA
- Al Nahda Tower, Sharjah 28 April 2012
- Mermoz Roubaix, France 15 May 2012
- Polat Tower, Istanbul, Turkey, 17 July 2012
- Tamweel Tower, Dubai, 18 November 2012

- The Torch Dubai February 2015
- Azerbaijan May 2015
- The Address. Dubai December 2015

I was taken aback when I became aware of this presentation. Even as late as 2016 I would have expected to have seen this matter highlighted. Yet I cannot recall any training or information regarding cladding fires and the importance of considering changing the stay-put policy, let alone details of *how* we were to evacuate, given our high-rise procedures.

Summary

On 13 June 2017, the day before the fire, and seen by the civil servant responsible, the Fire Sector Federation reported Approved Document B was 'out of date, placing businesses and communities all over the UK at potentially fatal risk'.[58] Also, the day before the fire, the TMO at RBKC had 287 actions outstanding (128 more than a year old).[59] We see here two major strands leading to this point, the second being the deterioration in standards of general fire safety, maintenance and inspections in residential high-rise blocks, accelerated by the introduction of ill-defined, unqualified fire risk assessors under the The Regulatory Reform (Fire Safety) Order 2005. This is often very visible to residents and firefighters conducting familiarisation visits.

Far more serious though is the strand that is hidden: a dumbing down of testing, certification and building control can be hidden in the very walls. How can a building control officer know a test certificate is not worth the paper it is written on? Once a refurbishment is complete, the lack of cavity barriers and flammable material is hidden from view. How can a firefighter know that compartmentation will fail in more than one area at the same time?

On the evening of 13 June, Red Watch reported for duty. They completed their vehicle checks and ensured their breathing apparatus had enough air. Cylinders were changed, appliance water tanks topped up and hose counted. The one type of job they were well prepared for was, ironically, a high-rise. But not this time.

The last part of this book will cover the fire and aftermath. I would ask the readers to put themselves in the shoes of the first crews. You now know what type of procedures firefighters apply to such an incident. You know how many appliances and personnel arrive. You would not know, as they did not, any of the hidden dangers lurking in that building.

Part III

The Fire and Aftermath

Before looking at the details of the incident, it is worth noting one important point regarding the question of evacuation. The Inquiry phase one report found 'That decision could and should have been made between 01.30 and 01.50 and would be likely to have resulted in fewer fatalities'.[1] I had looked through the timeline prior to the report being made public. I had expected them to point to a slightly earlier and narrower 'window of opportunity', perhaps between 01:20 and 01:30. Yet the expert witness noted the time of 01:40 as a critical point, after which part of the stairs and lobbies became affected by smoke. I am not claiming I was correct. Indeed, I can see why the inquiry came to that conclusion. However, it does highlight that working out when an evacuation can take place safely is not an exact science.

The reason why this is important is not just about identifying what went wrong and when at Grenfell. Nor is it solely about looking at what might have worked better. The more important question is the future. Because we are asking someone to make a decision regarding evacuation sometime after they arrive, which could be a significant time after the fire started, we are asking them to judge the life risk from staying put against the life risk from placing scores of people on a single staircase. And we are asking them to know that staircase is safe and *will continue to be so* for the duration of the evacuation.

I would ask the reader to bear these points in mind as we go through the timeline of the incident. Many of the timings and details can be found across various reports to the Grenfell Inquiry website, most notably the *London Fire Brigade Operational Response to Grenfell Tower*.

First call

The time was a little after ten to one on the morning of Wednesday 14 June 2017. On the fourth floor of Grenfell Tower a smoke alarm sounded in the kitchen of flat 16. The occupier, Mr Behailu Kebede, was asleep on a mattress in the living room, having fallen asleep a little over an hour before. He had returned home from his job as an Uber driver around 11:30pm. Two women, Almaz and Elsa, occupied a bedroom each and were both asleep.

Mr Kebede thought the sound of the alarm was coming from the kitchen. When he entered the kitchen, he saw light-coloured smoke around the fridge-freezer and window area. The window was open about ten inches and the smoke was described as rising upwards from the floor level and coming towards him. He returned to the living room and grabbed his phone and a pair of trousers. He banged on the doors of his two housemates and started to shout 'fire!'. Going out into the communal area he alerted his neighbours, banging on their doors and again shouting 'fire!'.

He made the first call to the emergency services. This is logged at 00:54:29 precisely. Mr Kebede had the presence of mind to turn off the electricity main switch at the fusebox by the front door. Importantly, he also had the good sense to close the front door behind him. In doing so, Mr Kebede prevented smoke from entering the lobby.

Both his lodgers got themselves to the stairs, one having grabbed a large suitcase, an act that would later excite conspiracy theorists. Once outside, Mr Kebede received a return call from the emergency services where he confirmed that he had already called the Fire Brigade and they had in fact just arrived. Mr Kebede was seen on CCTV leaving the main entrance at approximately 00:59. The first fire engine arrived at 00:59:24.

So, in brief, Mr Kebede was woken by a fire alarm, discovered a fire, raised the alarm, ensured everyone got out of his flat and then raised the alarm with everyone on his floor, even though, in theory they should not have needed to leave their flats at all unless they were affected by fire or smoke. Mr Kebede did everything that any reasonable person would, given the same circumstances. The inquiry quite rightly praised his actions.

The fridge-freezer

The fire appears to have begun near the base of the fridge-freezer, possibly due to overheating in improperly crimped wires. The particular machine had no apparent faults. The fridge itself was about five years old. One likely cause was overheated wiring in a small box at the base of the appliance which caught the plastic insulation backing the fridge alight (interestingly US fridges are required to have a metal backing).[2] Whilst the arguments about the exact component or piece of wiring may continue, the location of the seat of fire was known fairly early on in the subsequent investigation. Around the rear of the base of the fridge near the window of the kitchen.

It might well have turned out to be just one of the many kitchen appliance fires that occur each year (some 16,000 between 2012-18).[3] We all have many electrical devices in our homes. I've just counted twenty-two downstairs in my own home with a likely similar number upstairs. From kettles, TVs, PlayStation to mobile-phone chargers. If we multiply this by the over 26 million dwellings in the UK, then we have hundreds of millions of electronic devices just in the home. Possibly over a billion. A small number will malfunction. In the vast majority of cases, the item will simply stop working. Occasionally, a fuse will trip, thus preventing the device from overheating. A tiny percentage will malfunction *and* cause a fire, often either from overheating or causing a spark.

However, a tiny percentage of a very large number means fires caused by electrical appliances happen every day. Which is why you should not leave things plugged in that don't have to be, especially overnight. We can't help leaving fridges and freezers on, but I always turn things like phone chargers off overnight.

Initial indications and subsequent investigations point to this being a routine kitchen appliance fire. It's exactly what Mr Kebede described and exactly what the first crew was confronted with. After the fire there was a flurry of conspiracy theories about exploding fridges, bomb-making factories and many others. There's no evidence for any of that. It may stem from one part of a slightly inaudible call Mr Kebede took from the operator outside the building just as the Brigade arrived:

Mr Kebede: Allo ha.
Operator: Are you out?
Mr Kebede: I call… explode
Operator: What's on fire? What's the fire?
Mr Kebede: It's actually started in the fridge
Operator: In the fridge?
Mr Kebede: Yeah

In his statement he says he first saw smoke coming from the area of the fridge and window. He said the same thing to the firefighters on arrival. It was when he was outside, he stated, that he first saw flames coming out of the window. A rapid escalation of a fire isn't an explosion, any more than loud crackling or popping noises from a fire are indications of an explosion. Yet I have heard countless members of the public use such terms at many other incidents. The fact is the fridge didn't explode. Pictures of the kitchen and fridge freezer can be seen on the inquiry website as can all the expert reports.

Before we move on from the fridge, let us turn to the experts. Professor Niamh Nic Daéid found that 'The fire which occurred in flat 16 Grenfell Tower on the 14th June 2017 started in the kitchen of the flat. On the basis of the available evidence, it is more likely than not that the area of origin of the fire was in the tall fridge-freezer in the southeast part of the kitchen.' It was found that the fire extended out of the kitchen window. Later the fire re-entered the flat via the window of the bedroom next to the living room. This subsequently caused a flash-over and the fire spread into the corridor and back into the kitchen.

The electrical examination of materials recovered from flat 16 was undertaken by Dr Duncan Glover, a forensic electrical engineer, who reported to the inquiry. He found that two circuit breakers had tripped prior to Mr Kebede turning off the main switch. These were the individual circuit breaker for circuit number 7, the kitchen; and the residual current circuit breaker (RCCFi), which covered two circuits, number 7 the kitchen and circuit number 8 the flat sockets. He thus concluded the fire origin was within circuit 7. He concluded, 'the most probable fire origin is within BPS/1, the fridge-freezer, Hotpoint Model FF175BP.'

For those readers interested enough one can visit the inquiry website and view the pictures. The laminate flooring of the kitchen has a burned area directly below where the fridge-freezer stood and the skirting board directly behind the appliance was burnt away.

In conclusion, the occupier reported a fire emanating from the area where the appliance was standing. The initial crew, as we shall see, were confronted with a fire in that same area. Fire investigators would later identify this area as the seat of the fire. Experts to the inquiry would support this conclusion. The fire damage to the kitchen can be seen by those who care to look. But lies travel quicker than the truth and speculation fills the void when information is lacking or slow.

As early as the next day fire investigators were already sifting through the evidence, a painstaking process that can take weeks with even a small fire. A

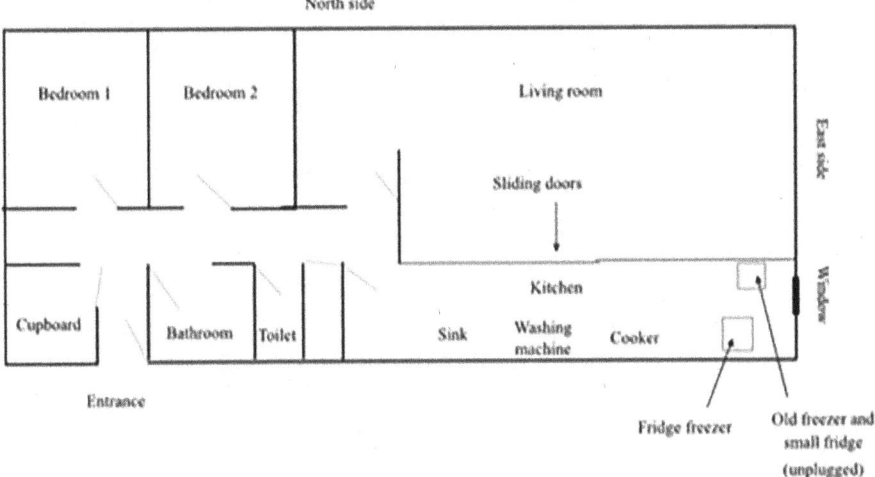

Figure 5: Flat 16 layout.

process that requires a scientific approach and patience. No such concerns for conspiracy theorists, several of whom had already got their running shoes on and were busy spreading their nonsense as the morning dawned and a shocked public scrambled for answers.

Suitcases, fridges and conspiracies

On his way out down the stairs Mr Kebede noticed one of his flatmates, Almaz, had a bag with her. He remembered it enough to remark on it in his statement describing it as a 'large carry-on bag'. Almaz described it as a bag she kept clothes in. She grabbed it on the way out and carried it to the top of the stairs leading out. Not an unreasonable thing to do if it was there to hand. CCTV caught images of Almaz on these stairs at 01:03. It certainly looks like a suitcase. She was there for some time. Firefighters passed her en route to the fire floor. A little later, a man, presumably a resident, helped her carry it down the stairs. The time stamp is 01:14:08.

Claims that Mr Kebede had carried a suitcase are completely unfounded. Claims that several people had time to pack and that whole families lugged several suitcases down the stairs are equally false. What appears to have happened is that one person grabbed a suitcase. She may have been storing clothes there and grabbed it quickly. She may have grabbed as much as she could in panic and stuffed the suitcase full before leaving.

Not the advice firefighters would give as we always say 'get out and stay out'; do not stop to grab possessions. However, Ms Almaz is not be the first to grab a few things before leaving and no doubt won't be the last. Not everyone thinks straight when they are panicking. Some don't panic at all, even when a little urgency is called for. None of this justifies the ridiculous claims and conspiracies about Mr Kebede or any of the other residents.

A second freezer and small fridge that were present in the kitchen were not plugged in and played no part in the cause of the fire. The older freezer had stopped working about nine years previously and Mr Kebede had never got round to fixing or getting rid of it. The small fridge had been purchased as a stopgap a few years before when the main fridge-freezer was being defrosted. Once that was working again, the smaller fridge had been unplugged and left on top of the old freezer. There they sat waiting for conspiracy theorists to concoct tall tales about why a man has two fridges.

First arrivals

We recall Mr Kebede called 999 and reported the fire at 00:54:29. The call, and the operator's response to that call, took just 45 seconds. At 00:55:14 a control officer in Fire Brigade Control despatched three fire engines: North Kensington's Pump and Pump Ladder, G272 and G271, and Kensington's Pump Ladder, G331. They were riding with five, five and six firefighters respectively, giving a total of sixteen.

Whilst en route control received two further calls to the incident. Watch Manager Dowden, riding in charge of G271, informed control that the address was a high rise and requested a further appliance. Control added a fourth appliance to the incident, Hammersmith's Pump, G362.

Mr Kebede and others left the building at 00:59 in time to see the first appliance, North Kensington's Pump Ladder, G271, arrive at 00:59:24. This is the time the driver or officer in charge presses 'status three' (meaning arrived) on the appliance console. It could be just as the appliance pulls up or a few seconds later. Likewise, one is supposed to press status two, meaning 'en route', as soon as one leaves the station on a call. If the button for status two was pressed on G271 it didn't register and control had to confirm via radio that they were in fact en route. As it turned out, they were the first appliance to arrive, so the exact time of logging status two is not relevant.

The appliance did press status three on arrival. It's worth noting that it is not always an accurate indication of the exact time arrived. I've sometimes pressed it as we've pulled into the road, usually because I'm worried I'll forget later. It's also slipped my mind and I've had to get my driver to check a couple of minutes after. Occasionally, people forget altogether or believed they pressed it but it did not register. Either way, if you didn't press the status button correctly one would receive a sharp rebuke by email or from the station commander. Fat fingers are no excuse.

However, on this occasion it would appear buttons were pressed correctly and the arrival time is accurate. Thirty seconds later Kensington's Pump G272 arrived.

Table 3: First appliances despatched and arrival times.

Appliance	Riders	Despatched	Status 2	Arrived
G271 North Kensington PL	5	00:55:14	00:58:48*	00:59:28
G272 North Kensington P	5	00:55:14	00:56:52	00:59:24
G331 Kensington PL	6	00:55:14	00:56:46	01:08:33
G362 Hammersmith Pump	4	00:59:12	01:01:16	01:08:27

* G271 booked status two via radio en route; this did not affect their arrival time.

A few words about this initial mobilisation. The pre-determined attendance for a dwelling fire at this time was three appliances. However, a residential high-rise should have had four. In the past an aerial appliance, such as a turntable ladder or hydraulic platform, would have been sent as well. However, at the time of the fire this wasn't the case.

Personally, I'd prefer an aerial on a high rise. But it's worth remembering that high-rise fires are fought from the inside. The building is designed to keep water out. If the window has failed, then those waiting to be rescued or firefighters searching won't appreciate a high-powered jet being delivered from outside into the same room. Aside from the force of the jet of water debris would be thrown about the room. In addition, there is the potential for a rapid increase in steam which could be fatal for someone in the room at the time. There's only so much an aerial appliance can do. It doesn't have an infinite reach. It requires enough space to manoeuvre into position and deploy its jacks. In general, I'd rather have one than not but it's not a magic bullet.

Regarding the attendance times, we can see from the table above that the first two appliances arrived within four minutes of being despatched and within five minutes of Mr Kebede's first call to the Brigade. They arrived as Mr Kebede exited the building and was on the phone again to the emergency services operator. The attendance times therefore seem reasonable and have no bearing on later events.

As we can see from all the above, with two appliances and a minimum crew of eight, you've got a hectic first few minutes to get everything in place before the breathing-apparatus crew can start up their sets and get their hose charged with water. At Grenfell they had ten and so initial numbers on arrival were not a problem.

However, once everything is in place and the breathing-apparatus crew start to gain access, it is preferable to have a back-up crew with another jet. This requires another appliance. It is also policy to have an emergency crew as soon as possible. Hence four appliances is about right for a domestic high-rise fire following procedures in London. In summary, the arrival times, number of appliances and number of firefighters were not a significant factor at Grenfell.

The first two appliances arrived at 00:59 and WM Dowden, now the incident commander, met Mr Kebede on the ground floor outside the south elevation. Mr Kebede told him the fire was on the fourth floor; he believed it started in the fridge-freezer in the kitchen and that everyone had got out of the flat. The driver of North Kensington's pump located the hydrant and began laying out two lengths of 70mm hose whilst a firefighter laid out hose to the dry riser main inlet at the base of the tower.

The incident commander ordered Crew Manager Secrett to set up a bridgehead and commit a breathing-apparatus team once they had water. Meanwhile, people were already leaving the building. Occupants of flats 14 and 15 from the fourth floor, but also a Mrs Alves from flat 105 on the thirteenth floor.

At 01:00:28 the first station manager was paged by Brigade Control and notified of the incident and the fact that there had been further calls. He subsequently called brigade control and decided to monitor the incident and wait for the first information message.

Crew Manager Secrett could not access the ground-floor lift lobby as it required a key fob. Luckily, Mrs Alves was close by and allowed him to use hers. This lost a few seconds but, at 01:01:16, CM Secrett entered the lift lobby on the ground floor. He was accompanied by CM Batterbee and FF Brown wearing breathing apparatus (not turned on at this point), each carrying a 45mm length of hose. Three other firefighters accompanied them, carrying an Entry Control Board, thermal-image camera and hose. This left the officer in charge and pump operator outside, along with two other firefighters to assist with the hydrant, hose and dry riser.

One of the first problems encountered was the fact that the fire lift did not work. Turning a key in the control switch should have brought the lift car to the ground floor and allowed fire-fighters to take control from inside the lift car. This would have disabled all the lobby buttons effectively, preventing residents from using the lifts and aiding firefighters in carrying equipment to, and assisting rescues from, the bridgehead. Instead, the lift had to be called manually and remained available for residents to call it from any of the lobby buttons.

One of the experts reporting to the inquiry found that the lifts were not compliant with the standards for firefighting lifts.[4] Firefighters were unable to take control of the lift as features were 'defective and did not operate as intended'. In addition, residents were still able to use the lift during the incident. Afterwards the lift cars were found on the tenth floor. Whilst questions were raised about whether the correct key was used, the mechanism was subsequently found to have been jammed and contaminated with builders' debris.

Whilst this caused significant problems for firefighters during the incident the consequences for residents were graver. The magazine *Inside Housing* reported that 'three residents are believed to have died after the lift they were in stopped and filled with smoke on the 10th floor at around 1.25am'.[5]

Outside the tower, smoke and flames could be seen through the kitchen window of flat 16. Back at the lift, a firefighter entered the building and handed a firefighting branch and Immediate Emergency Care pack (a large first aid kit) to firefighters in the lift. At 01:02 five firefighters reached the second floor. Two others ascended to the second floor via the stairs.

By 01:03 the bridgehead had been established on the second floor and hose connected to the dry rising main on the ground floor. A minute later, three firefighters went to the fourth floor and set up a line of hose to attack the fire. At the bridgehead on the second floor, CM Batterbee and FF Brown donned their face masks and started up their sets, their tallies placed in the breathing-apparatus control board. Outside, WM Dowden and Mr Kebede walked along the south side of the building to the south-east corner of the block to get a better look.

The first breathing-apparatus crew, BA team 1, made their way upstairs. Outside, fire and smoke could be seen coming out of the window of the kitchen on the fourth floor. Someone asked for a covering jet just as the hose leading to the DRM inlet can be seen on CCTV filling with water.

By 01:06 Crew Manager Secrett at the bridgehead received confirmation that water was available. In that brief radio message, WM Dowden mentioned the covering jet and was advised not to direct a jet into the flat as the breathing-apparatus crew were about to enter the flat. Seconds later, BA team 1 had arrived at the flat door carrying a charged 45mm jet, thermal-image camera and enforcer (basically a hand-held battering ram). They gained entry and thick black smoke could be seen coming from the flat as they did so.

The three firefighters in the lobby of the fourth floor retreated into the stairwell and closed the door behind them to prevent smoke entering the stairwell. Outside on the ground floor, the fire could be seen breaking out of the kitchen window. The fire in the kitchen is described at that point as fully developed. This meant the fire had spread over most if not all the available fuel in the room. It is also described as venting, meaning smoke and heat were exiting the opening whilst clean air was being drawn in. A single open window will allow both to occur: heat and gases escaping through the top of the opening and air being drawn in at the bottom. So at this point it appears to be a well-developed fire but one that a crew with a 45mm jet should be able to extinguish.

Hammersmith's pump arrived and two firefighters in breathing apparatus were sent to the bridgehead. A few seconds later Kensington's pump ladder also booked status three. There were now four appliances and twenty firefighters in attendance. The fire at this point was a regular flat fire. A 'bread and butter job'. Nothing to indicate the horrors to come.

We can see the floor and flat numbers in figure 6. Figure 7 provides

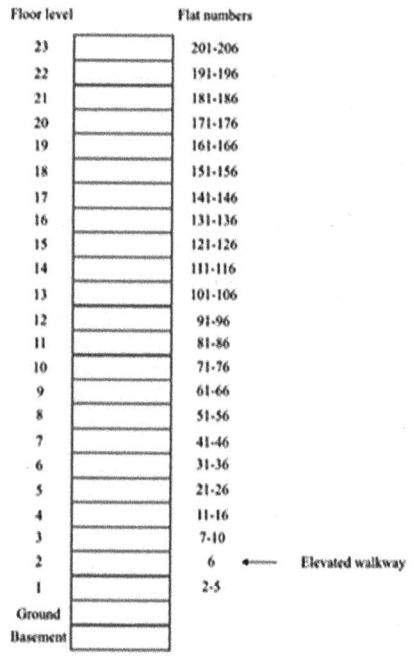

Figure 6: Floor and flat numbers.

a plan, or bird's eye view, of the fourth floor where the fire started in flat 16 in the north-east corner. Every floor above the fourth mirrored the same layout with four two-bedroom flats and two one-bedroom flats, numbers ending in 1 to 6, which can be seen in figure 8.

One firefighter had gone to the fifth floor, alone and without breathing apparatus, where he met a family who informed him that they had come

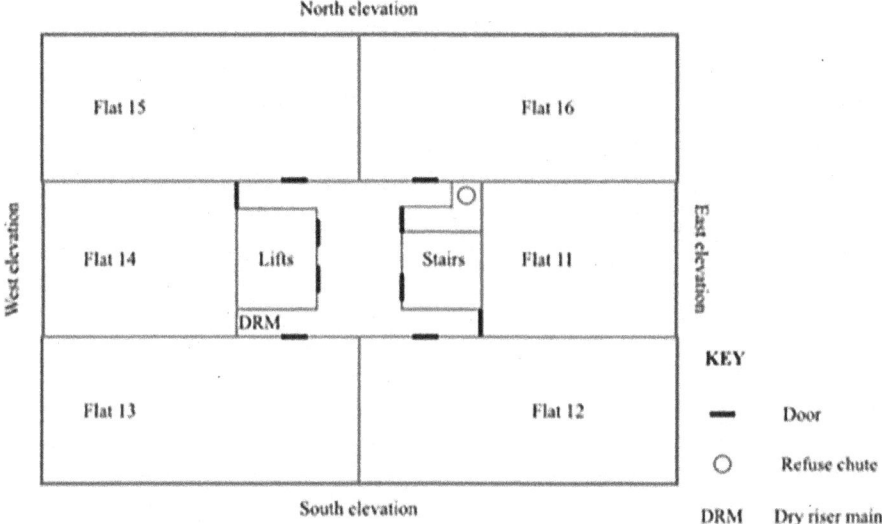

Figure 7: Plan of fourth floor.

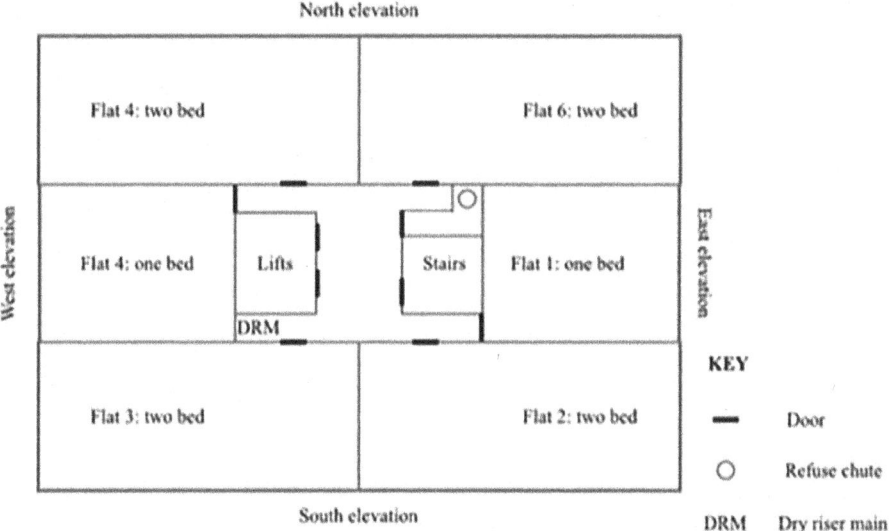

Figure 8: Plan for floors 4–23.

out of flat 26, directly above the fire. They told him their flat was on fire. However, when he checked the door, it was locked, so he looked through the letterbox. The light was on, but there was no sign of smoke let alone fire.

Meanwhile BA team 1 was searching the bedrooms in the flat, standard procedure even if told that no one is involved. Outside, flames started exiting the window and embers could be seen falling to the ground. Within two minutes, two further breathing-apparatus crews were making their way to the bridgehead and a second jet was set into the dry rising main on the third floor. A covering jet on the ground floor can be seen filling with water on the east side. This was below the window of the kitchen in flat 16. BA team 2 reached the bridgehead and started up under air.

Up to this point in the incident it had been a fairly normal flat fire in a residential high-rise. The only indication that something might not be quite right came on the fifth floor. But when the firefighter checked there was no sign of fire.

First signs of disaster

At 01:12 the first visual indications of any problem appeared. The incident commander noticed that the external cladding just above the window of the kitchen on the fourth floor seemed to be burning. He described it as spitting and sparking, similar to magnesium. An immediate request for two further appliances was made, along with a hydraulic platform (an aerial appliance that can reach about 30 metres in height). It is worth bearing in mind that the tallest aerial could only reach the eleventh floor. Any fire behind cladding on the upper floors would be inaccessible to even the tallest aerial.

The first photographic evidence of the exterior cladding alight is timed at 01:14. The inquiry learned that any fire near a window was very likely to spread to the external rain-screen cavity. The majority of construction materials had no fire-resisting performance. In fact, the presence of combustible materials aided fire spread. Additionally, the windows lacked fire-resisting cavity barriers around the perimeters.[6] Once the fire penetrated the rain-screen cladding there was no provision in the system, neither materials nor their arrangement, to impede the spread of fire and smoke. This 'created the means for a catastrophic condition'.

Outside the firefighters were unaware of the built-in catastrophic failures. The request for further fire engines triggered a number of responses from control. They sent two further pumping appliances, two command units, a fire investigation officer, a group manager and four station managers.

Meanwhile, in flat 16 BA team 1 had searched both bedrooms and the toilet as they made their way towards the kitchen. A watch manager was sent to take over the bridgehead. More problems with access delayed him for several seconds. Shortly after BA team 1 reached the kitchen and opened the door for the first time. They could see no visible flame through the smoke and any water they applied turned to steam.

Those outside could see the flames from the kitchen clearly but inside several feet of thick smoke was easily able to obscure this. Visibility in the flat was described as zero. The thermal-image camera recorded images used in the inquiry. It showed a fire in the far corner of the kitchen above and around the fridge-freezer which can be made out on the image. Fire is also seen around the window area which can also be made out on the image.

The first informative message was sent: 'From G272 residential block of flats of 20 floors 25 metres x 25 metres, five roomed flat on fourth floor, 75 per cent alight, high-rise procedure implemented MDT in use, tactical mode Oscar.' This is a standard form of words used to give control basic details of the incident. Tactical mode Oscar simply means 'offensive', (i.e. firefighters are committed and actively attacking the fire as opposed to Delta, defensive, which means not committing firefighters but rather surrounding with covering jets.

At 01:15 a firefighter on the third floor met individuals exiting the building. They showed signs of having been in smoke: eyes streaming, coughing and looking panicked. Meanwhile, a firefighter on the fifth floor looked through the letterbox of flat 26, directly above flat 16. He saw the hallway of the flat was full of thick black smoke. The light that previously he could see left on was now obscured by smoke. He attempted to radio the incident commander but could not get through on his handheld radio.

Outside a large quantity of debris was falling externally. The covering jet appeared to have little effect, mainly because the fire seemed to be spreading behind the panelling. The external panelling is designed to keep rainwater out, so it's an impossible task to get water onto any fire behind it.

In the flat BA team 1 made repeated attempts to get into the kitchen but the jet had no effect on the fire. They closed the door and attempted to find another entrance via the lounge on their left. A minute later they were back at the door where they were met by BA team 2 in the hallway outside the kitchen.

At this point brigade control was still of the view this was a 'standard fire' when the first station manager was ordered on. Back at the bridgehead, the information had got back about possible fire spread. BA crew 3 went under air and were ordered to proceed to the fifth and sixth floors to check. Even at this point I would be inclined to think that the reports of fire was probably smoke spread via an open window from the flat below. We recall smoke and not flames were seen in flat 26.

By 01:18 a firefighter on the sixth floor was told by a family from flat 36 (two floors directly above flat 16) that their flat was on fire. He entered the flat and found a wall of thick black smoke. Again, there were no visible flames. Outside the firefighters had been desperately trying to extinguish what they saw as an external fire in the cladding. But, as the first and second panels caught fire, it began to grow rapidly up the side of the building.

The report by Barbara Lane marks 01:14 as the point at which fire started to spread up the building's east face. It then spread vertically, reaching floor 13 by 01:22. By 01:26 it had reached the top: nineteen stories in twelve minutes. Once within the cladding there were six different 'pathways' of fire spread.[7] The technical details of these can be seen on the inquiry website.[8] The result of these pathways allowed vertical and horizontal fire spread around the entire building exterior and can be seen in table 4.

Table 4: Pathways of fire spread.

Pathway	Details
A	Vertical spread up (and down) the full height of the columns.
B	Horizontal spread across the Reynobond spandrel panels.
C	Horizontally along the edges of the head and sill of the windows, and the edges of the top and bottom of the insulating core panel.
D	Vertically along the window edge and along the edge of the Aluglaze insulating core panel.
E	Vertically by means of the Aluglaze insulating core panels which connect between spandrel panels.
F	Around the crown of the building façade.

Of course, on the night firefighters outside had no idea about the details of the mechanism of fire spread. All they saw was a fire rapidly spreading externally. No surprise if smoke had somehow got back into some of the flats above via open windows. What they could not possibly have known was how exactly the fire had spread from the kitchen to the rear of the panel system.

Even more important was the fact that the problems with the windows were not confined to the kitchen window of flat 16. The general standard of the construction and fitting of the window and its surrounds, the lack of cavity barriers and flammable filler would allow fire to spread back into the flats above. In fact, it had already done so on several floors vertically above the original fire.

Internally, the person in charge at the bridgehead had done exactly as expected and sent a crew to investigate. A priority message to 'Make Pumps 8' had been sent at 01:19. By this time smoke had spread into the lobbies on the fifth, sixth, fifteenth and sixteenth floors. It is likely at this point that defective door closers and fire doors were already having an effect. The table below shows the subsequent rapid requests for more resources. This is very unusual and shows the dynamic nature of the incident.

Table 5: Initial assistance messages.

Time	
01:13	Make Pumps 6 and Aerials one.
01:19	Make Pumps 8.
01:27	Make Pumps 15, aerials x 2.
01:29	Make Pumps 20, FRU x 2.
01:31	Make Pumps 25.

At 01:20 BA team 3, sent to investigate smoke in a flat above the fire, found heavy smoke logging on the fifth floor. They had no breaking-in equipment and were unable to access flat 26 but they could see smoke inside. Back in the original fire compartment, BA team 1 opened the kitchen door again and noticed a drop in temperature, indicating the fire was 'venting'. They entered the kitchen and extinguished the fire. Externally at this point the fire had spread up the east elevation to the ninth or tenth floor. Within thirty seconds it had spread to the eleventh floor.

A fourth breathing-apparatus team was committed to extinguishing a potential fire on the fifth floor. Meanwhile, a firefighter reported by radio that fire had spread to the seventh floor and another found smoke logging in the lobby of the eighth. In fact, fire had spread to eight flats by 01:21 and twenty flats by 01:26.

Back in the original fire compartment on the fourth floor BA team 1 had been attempting to extinguish what they could of the external fire. One firefighter was leaning out of the window directing the jet upwards whilst being held by his colleague. By 01:23 the fire had reached the twelfth floor and just two minutes later the twentieth.

From the early stages falling debris was a constant hazard to both firefighters and those evacuating. The effectiveness of aerials was hampered. Spotters were required to assist firefighters when it was safe to dash across the 'danger zone'. At one stage, an alternative entry was made by breaking a large glass pane to allow access into the ground floor. Firefighters had to put out developing fires on the ground floor. In addition, fire hose feeding the dry riser inlet was struck and burst and required replacing. Police shields were used to protect firefighters as they entered and exited the building.

As the fire raced up the cladding system, the gap between the insulation and panels allowed a chimney effect. The insulation aided combustion. The lack of cavity barriers and fire stopping allowed fire spread. Any exposed PE core in the panels caught quickly. If the rear metal sheet of a panel peeled away due to the heat, more of the flammable core was exposed. This caught quickly and added to the rapidly escalating fire.

All the time, smoke and flames were re-entering flats either through open windows or the gaps surrounding them. Any flammable filler around windows burned quickly and UPVC melted away. Rooms quickly became involved as the fire jumped from floor to floor.

So, by 01:26, fire had spread externally up to the roof level and internally to twenty flats. The initial attendance was designed to deal with a fire in one flat. It can just about deal with fire spread to another flat. Nor is the dry riser capable of delivering sufficient water pressure to supply a firefighting jet on every floor at the same time. To deal with twenty flats and an external fire covering an entire face of a building would need twenty appliances just to provide one breathing-apparatus team per flat. In the event a fifth and sixth appliance arrived by 01:26.

Barbara Lane told the inquiry that, the 'information allows me to conclude that the principles of the stay put regime can be considered to have started to fail by 01:15 (time fire spread to Level 5), and to have substantially failed by 01:26 (fire had spread to Level 23)'.[9] Additionally, 'Therefore, the primary consequence of the rain-screen cladding fire starting at Level 4, and spreading seven storeys within 7 minutes, and 19 storeys within 12 minutes, was that it rendered the Stay Put strategy unfit for purpose before 01:26'.[10] The timeline up to this crucial time can be seen in table 6.

Table 6: Timeline 00:54-01:26.

Time	Comments
00:54	First call to the fire service to a fire in kitchen of flat 16, Grenfell Tower.
00:59	First two fire appliances arrive.
01:03	Crews enter 2nd floor lobby and set up bridgehead.
01:06–07	Crews arrive at front door of flat. Flames already penetrated window in several places.
01:08	Two further appliances arrive.
01:12	External cladding above window alight.
01:13	A 'Make pumps 6, aerial required' message sent to control requesting two additional appliances and an aerial appliance.
01:14	Crews access kitchen for 1st time. Externally fire extended two floors above flat 16 on 4th floor.
01:16	BA team sent to fifth floor to carry out reconnaissance.
01:17	Firefighters on 6th floor told by residents there was a fire in one of the flats.
01:18	35 residents had left the building
01:19	'Make pumps 8' message sent. FRU with extended duration breathing apparatus sent.
01:20	Crew in flat 16 enter kitchen and extinguish fire. Fifth floor heavily smoke logged.
01:21	External fire spread to 11th floor
01:22	Firefighter inside tower reports via radio fire on 7th floor.
01:23	Firefighter reports heavy smoke logging on 8th floor
01:24	'Make pumps 10' message sent
01:26	Fire had reached the 23rd floor. Expert witness Barbara Lane reported stay-put policy had 'substantially failed'.

Post 01:26

We now get to the point where the inquiry found there was a window of opportunity. Shortly after this time, the Brigade had received twenty-three calls and a message was sent, 'Make pumps twenty and two Fire Rescue Units.' A male caller on the twenty-second floor stated the 'conditions are terrible and it is not possible to see your hand in front of you'.

As the crown became involved, the fire began to wrap around to the north face. A short while later the top twelve floors of the north-west corner of the north face became affected. Back on the east face the fire was also spreading laterally affecting the middle flats adjacent to flat 6 of each floor.

A 'Make Pumps twenty-five' message was sent just as the first aerial appliance, Paddington's turntable ladder, and the first station manager arrived. By then, there were several fire survival guidance calls and so the decision was made for the station manager to take charge of those rather than take over as officer in charge. By 01:36 Control had received thirteen fire survival guidance calls on floors 11, 12, 14, 18, 20, 22 and 23.

On the eighteenth floor a caller reported that she attempted to leave but was unable to due to 'thick black smoke', presumably in the lobby. At 01:33 BA team 3 entered the lobby on the sixteenth floor and found it smoke logged. They crawled in and come across Edward Daffern from flat 134, the very man who, on behalf of tenants, had warned the TMO about the standards of fire safety at Grenfell. They assisted him to the staircase from which he evacuated. The crew then entered flat 136, twelve floors directly above the original fire. They found conditions very hot and heavily smoke logged. With no water and low on air, they withdrew, knocking on the other flat doors as they went. They got no response.

A minute later a caller on the 22nd floor tried to get to the stairs but found the lobby 'full of smoke'. Callers on the eighteenth floor reported 'thick smoke' in their flat. On the twenty-second floor smoke was again reported in a flat and on the twentieth a fire. Around the same time, a breathing-apparatus crew reported thick smoke in one of the lower lobbies between floors 6 and 8.

At around 01:36 the crew manager of the first fire rescue unit found the incident commander. In a rather desperate move, he was asked to get a line operations system working from the roof to allow water to be put on the fire from above. The intention was for a drencher-type system to put the fire out from the outside. Fire was now spreading laterally to the corners of the east elevation.

At ground level, residents entered the block and informed firefighters that people were threatening to jump at the rear of the block. Some firefighters rushed round to investigate. On an upper floor a distressed caller with a baby reported black smoke in the corridor and that people had tried to get out but were unable to get through the smoke, which was now entering the flat. She was advised to block gaps and openings. Between 01:19 and

01:38 ninety-six residents had self-evacuated, been assisted out or had been rescued. Brigade control had received sixty-seven calls.

More were to come. From the top floor a report of a fire on the floor below and that six people were in flat 205. On the fourteenth floor, smoke and fire was 'coming through the door'. On the twelfth a fire was in next door's kitchen. A caller from flat 182 on the twenty-first floor stated that they were trapped, having tried to get down the stairs but found it was 'too smokey'. At 01:39 a male caller, on his own, in flat 204 on the twenty-third floor reported a little smoke coming into his flat. He was asked if he could get out and he replied that he could not.

Here then we have several examples of people being unable to get out even when it is suggested. That indicates just how bad conditions were in several of the lobbies. On the stairs a crew between the fifth and sixth floors had already reported the stairwell filling up with smoke as early as 01:38. In addition, hose lines were beginning to clog the stairs.

The smoke ventilation system was also having an effect on the fire. The inquiry heard that 'the existing design of the tower had made designing a compliant system impossible'.[11] An expert witness to the inquiry wrote that the smoke control system 'failed to provide a smoke ventilation system which complied with the functional requirements of the building regulations.'

The system was designed to work when only the doors to the stairwell doors and none of the flat entrance doors were open. This on its own was an unrealistic assumption. Even in the best-case scenario, somebody might open a door. At Grenfell multiple people were forced out of their flats as the fire initially raced vertically upwards. There was hardly a moment when one or more doors were not open. The failure of door self-closers and fire doors failing made the situation far worse. Witnesses saw smoke leaking out of the vents on the sixth, twentieth and twenty-third floors from as early as twenty minutes into the fire.

Meanwhile, back at 01:26, one of the initial crew to arrive, FF Badillo had met Ms Urbano at the main entrance. She handed him the keys to flat 176 on the twentieth floor where her sister Jessica was and he told her that he would get her. He attempted to reach her without breathing apparatus via the lift, but it stopped on the fifteenth floor and filled with black smoke. Fortunately, he was able to find the staircase and get down. As he was doing so, control received a call from a young girl on the twenty-third floor in flat 201. The girl's name was Jessica.

FF Badillo grabbed a breathing-apparatus set and two other firefighters, CM Secrett and FF Dorgu. At 01:33 they went under air and started for the twentieth floor where they believed Jessica still was. By 01:36 they were in the lift heading up, but it stopped unexpectedly on a lower floor. The doors opened and thick black smoke rolled in. The crew made for the staircase and from there the twentieth floor. In the ten minutes it took to reach the twentieth floor they met no one on the way down. When they reached the twentieth, they found the flat door open. They searched it twice but found it empty, attempted to radio but couldn't get through.

Suddenly, the temperature on the floor soared to being unbearable in both the flat and lobby. The crew barely made it to the stairwell. It was a close call with one firefighter curling up in a corner convinced he wouldn't make it. Somehow, they found each other and made their way down the stairs, which at that point were in parts heavily smoke logged. They made it to the bridgehead and closed their sets down at 01:57.

The time this crew were under air coincides with when the inquiry found the following: sometime between 01:40 and 02:00 conditions worsened considerably on the stairs.[12] Thick black smoke reduced visibility to zero. Lobbies on floors 6-10, 14, 19 and 20 were all 'smoke filled'. Those between 6 and 10 were even hotter than the stairs. Later, between 02:00 and 03:00, there was thick black smoke between floors 7 and 12. The tenth floor was described as 'boiling hot'.

The inquiry was told: 'In general, from 0:55 to 01:30 the stairs appear to have been free of smoke and therefore tenable for escape' but significant smoke logging occurred after 01:40.[13] It acknowledged this information was not available to decision-makers. The time between roughly 01:20 and 01:40 saw the greatest number of people evacuating. Residents of every floor escaped during this time, except Levels 4, 22 and 23. At this point 151 residents remained in the building. The reader will note that one might interpret this as at odds with the phase one report which pointed to a window of opportunity between 01:30 and 01:50.

From 01:40 the situation got steadily worse. A caller on the fourteenth floor reported that the whole flat was full of smoke, and he was trapped in his bathroom. On the twenty-third floor another flat was full of smoke and on the tenth the caller stated he could not get out. The door was hot to the touch and black smoke filling the lobby outside was coming in. Outside a firefighter was directed to get a loudhailer and re-assure people threatening to jump. By 01:41 there were ten fire engines, one command unit, a fire rescue unit and a turntable ladder in attendance.

With fires reported on so many floors, it is worth noting that dry and wet risers are designed to supply 1,500 litres a minute.[14] This would allow three jets at 450 litres per minute each. The riser, or indeed the entire building, is simply not designed to cope with fires in multiple flats at the same time. At a push, one might supply six adequate jets, although it requires the higher flow rates to reduce chances of flash-overs and back-draughts in compartments. What can't be done is supply enough water via a dry rising main to fight fires in dozens of flats at the same time.

Meanwhile, crews were attempting to evacuate floors. At 01:44 a breathing-apparatus crew started banging on doors on the sixth floor. They split up to take a couple and child to safety. Other residents shortly followed from this floor. Around this time a caller on the twelfth reported a fire next door and smoke in her flat. She had two children with her. A BA team on the fifth floor described the smoke logging as 'crazy' with the floor hot underneath and 'you couldn't really see in front of you in that lobby area'.

By 01:50 one hundred emergency calls had been made, ten breathing-apparatus crews had been committed and the twentieth appliance had arrived. The problem, however, was not manpower. It was how to get to people once the staircase had been compromised? In addition, how to get them through the lobbies, and down several flights of stairs, that were, in places, filled with hot, thick black smoke?

Breathing-apparatus crews began to be told not to don their face masks until they were further up the stairs to conserve air, going completely against procedures. One commentator noted the 'bravery of individual firefighters who bent, broke and threw away rules that would have hindered their rescue efforts' in addition to the 'moral courage of those commanders who approved and supported many of those decisions', knowing full well the potential legal consequences should these actions result in firefighter or resident fatalities.[15]

Meanwhile both BA teams 12 and 13 only reached the fourth floor before conditions on the staircase were so bad that they had to go under air. Despite this between 01:39 and 01:58 twenty people had been rescued, assisted out or self-evacuated.

By 02:00 rescues were being attempted on several floors.[16] Before this time there is evidence of the dry rising main being used successfully up to Level 9.[17] Beyond that water pressure was very poor.

At times firefighters, deeming the stairway too dangerous, were moving residents to a flat clear of smoke and considered safe. They would then inform entry control of their location so that other crews could lead them to safety. The flame front was spreading diagonally across the east and north faces. All this time fire was entering more flats via the gap between the windows and the original wall. With each faulty door self-closer or failed fire door, more smoke and heat was entering lobbies. As residents and firefighters exited and entered the stairs, this smoke and heat was drawn into the stairway.

A woman and daughter on the ninth floor waited with a breathing-apparatus crew for spare sets to be brought up for them to use. Both were led to safety. Two firefighters rescued two people from the nineteenth floor. Three firefighters returning from the upper floors came across a crew on the tenth struggling with an unconscious male and conscious female and assisted them. Two rescues were made from the fifth floor using a 13.5-metre ladder pitched from the walkway.

At this time 129 people were still in the building: twenty-nine on the twenty-third floor, fourteen on the twenty-second, nine on the twenty-first, nine on the twentieth, two on the nineteenth and at least eight on the eighteenth.[18] A further twenty-five calls were made to control between 02:00 and 02:20 including eleven fire survival guidance calls.

Soon after 02:00 the fire had spread laterally so that all flats on the upper floors of the east face began to be affected. Shortly after a message was sent making pumps forty. The fire on the north face began to spread both laterally and downward. Within half an hour most of the flats on the upper floors of the east and north faces were alight and the fire now began to wrap round to the south face.

Around this time a police control room operator spoke with a fire control room officer and asked what advice they could give to callers. It was explained that some callers had tried to leave but could not, due to the smoke but 'the Brigade do not generally tell people to leave but if they think they can leave safely then they should do so'.

On the tenth floor a crew tried to crawl into the lobby on their bellies. The temperature was 'extreme'. and they felt themselves 'burning' even through their protective fire gear. They were physically unable to push through the heat and unable to reach a flat. They had no water, but it is doubtful that if they did have that the pressure would have been sufficient.

A caller trapped in a kitchen on the top floor reported that they could not get out as the corridor was full of smoke. The flat was full of thick black smoke and the fire was now in the bedroom. The operator told them they had to decide 'but if the fire is in the flat then they need to get out' and by 02:30 the caller said they have to go, and the line went dead.

Several other callers reported that they attempted to leave but had to turn back. The front doors were hot. Some of the metal handles would have been too hot to touch. Lobbies were filled with thick, black smoke. Some were scorching hot. A wall of heat as soon as you opened the door. Those that reached the stairs hoping for clear air would have been confronted with more smoke and a huge dilemma.

Control officers had similar dilemmas. The advice was stay put unless you are affected by the fire or smoke. More and more they were emphasising this caveat. Soon after 02:15 someone at control had found a television in the room below and turned on Sky News. It was the first time someone in control had seen images of the fire. At 02:30 a caller was told that if the fire was in the flat, then they needed to get out. Another operator at 02:43 was told by the caller that the front door was hot and there was a lot of smoke. They did not think they could get out. They were told to find a room that was safe but that the advice was that they should get out if they could.

Another resident attempted to leave but had to turn back. They were advised to put wet towels around themselves and the doors if it was definitely not safe to get out and that crews were getting to people as quickly as they could. Two frantic calls from the top floors around 02:43 revealed conditions were such that they could not get out. They were advised to get out if they could. Four different callers were advised to cover themselves in wet towels and attempt to get out.

One firefighter who attended in these early hours described the experience:

We were called on in the initial stages of the fire in one of the early make ups. En route we could hear fire survival calls coming in and it was clear this was a major incident. On arrival several floors on two sides of the block were well alight and my initial feeling was the building was already lost. Members of the public were shouting about people trapped and we quickly took our breathing-apparatus sets and arrived at the south side of the tower. Debris was already falling from the tower.

We were given an initial task but were redirected to an upper floor. We encountered heavy smoke logging which quickly made reading the floor numbers impossible. By the time we reached the upper floors we were exhausted as we were also carrying equipment such as hose and breaking in gear. Some of this we discarded so we could continue on.

Close to our objective we came across residents who were in great distress. We had to decide quickly whether to let them go alone and continue to our original objective or change our brief. It was unlikely the residents would make it downstairs without help so we decided to assist them down the stairs although the amount of hose and equipment caused some problems. The smoke logging and lack of visible floor numbers also hampered us. Our decision turned out to be right as one casualty collapsed on the way. The bridgehead was chaotic with lots of firefighters receiving briefings or coming out after a 'wear'*. When we got down to the ground floor we could hear the fire roaring and loud bangs as bits of cladding fell off the building and crashed to the ground.

Once outside we went to replenish the cylinders on our breathing sets. Looking back at the tower I could see it was now completely ablaze, yet I could see a resident standing calmly in one window.

The duty Assistant Commissioner, Andy Roe, had arrived at 02:31 by which time 100 fire survival guidance calls had been made. Prosser and Taylor describe an 'element of confusion' around whether the stay-put policy had been revoked.[19] A watch manager on the command unit had told DAC O'Loughlin that control had stated the policy had been changed. This led him to believe residents were now being told to leave *if it was safe to do so*.

It does not follow from this that all residents who had made FSG calls or called after this time would be told to leave or were able to do so. If the conditions outside their flats were such that they could not leave, then they were trapped. If the handle or door is red hot you might feel reluctant to open it. If you do so and are hit with a wall of red hot thick black smoke, you might shut it quickly and decide you cannot go into the lobby.

In the event the official decision to revoke stay-put was made at 02:47. In reality, a de-facto decision had already been made at control ten minutes earlier.[20] At 02:50 an operator told a caller, 'if you don't do what I tell you,

---

* A 'wear' refers to a period of wearing breathing apparatus in a fire, typically of 30 minutes' duration.

you are going to die in that flat'. By that time firefighters were having great difficulties getting above the fifteenth floor.

BA Team 31 described conditions on the way up the stairs as deteriorating dramatically beyond the seventh floor: thick, heavy smoke with very poor visibility. A number of residents passed them coming down the stairs, some assisted by firefighters. None appeared to be slumped or unconscious. Between 02:30-3:00 thirteen people were rescued, assisted out or self-evacuated.

At 03:00 the east and north face was a raging inferno with flats on the top ten floors of the east corner of the south side well alight. The fire had also wrapped round to the west face of the building and began its unhindered spread laterally and downwards. Rescues continued to be made on upper floors but on the eighth a woman and child were found on the stairway. Both were taken out by breathing-apparatus crews but sadly did not regain consciousness.

The eleventh, twelfth and thirteenth floors were cleared of residents. By then conditions on the staircase had deteriorated to the extent that it seemed only those in breathing apparatus could use it.[21] Visibility was reduced by the thick black smoke. It seems the fluctuations of the fire and opening of doors by firefighters and residents meant conditions in the stairway changed back and forth. Two women were brought down the staircase from the twelfth floor and survived. A man found on the tenth and taken out did not.

A crew on the fifteenth floor witnessed smoke and flames flickering down the corridor, their thermal-image camera registering 550 to 555 degrees centigrade. Between 02:50 and 03:00 London control-room operators received over a dozen calls from inside the building. Each one was advised to get out if they could.

The Commissioner arrived at 02:50, three minutes after stay-put had officially been revoked. A tactical co-ordinating group meeting at 03:20 determined that at least 100 people were still trapped in the building. Breathing-apparatus crews continued to rescue people: two children and three women from the twenty-second floor, had been found on the ninth and brought down.

At around 03:00 the fire had affected all the flats on the east and north face of the top ten floors and a fair proportion of the flats below. It had also spread to the west and south faces. Of the latter, the top four flats on the east corner of the south face were well-alight.

At 03:13 BA team 33 attempted to clear flats on the ninth floor and registered over 1,000 degrees centigrade on their thermal-image camera.

The heat was intense and prevented repeated attempts to gain access to flats. Returning to the stairwell to get some breaking-in equipment, they discovered two adults and two children struggling to get down the stairs. They were able to assist them to safety. Meanwhile, smoke on the second-floor lobby had caused the bridgehead to be moved to the ground floor by 03:30.

By then smoke on the stairway was significant from the third to top floors but the conditions between levels 13 and 16 were particularly hot, at one point melting the plastic lights. Smoke-spread into the stairwell appears to have been caused by the frequent opening of doors by firefighters and residents. It is hard to see how that could have been avoided, given the circumstances. Despite the conditions, around 03:47 two residents from the sixteenth floor managed to come down the stairs safely and reach safe air. Yet another two casualties sadly did not and remained on the stairway around the ninth floor for some time.

At approximately 03:55 twenty-two of the twenty-four flats on floors 13 to 16 were alight.[22] Around 04:00 contact above the fourteenth floor ceased. Shortly after this time two women were brought to safety from the tenth-floor lobby. However, a woman found in the eleventh-floor lobby sadly died. A crew of four, sent to the eleventh floor to rescue a woman and child, was beaten back by the heat. Following guidance from control, they left their flat and entered the smoke-filled lobby. Later, at 04:27 BA Team 47 were met with an intense heat barrier and were unable to enter the eleventh-floor lobby. Fortunately, the woman and child attempted to escape, were seen by firefighters and guided to safety at around 04:47.

About a quarter of an hour after BA team 47 had been beaten back by the heat another crew managed to get through the lobby and rescue a man, woman and child. Between 04:12 and 04:47 nine persons were rescued, assisted out or self-evacuated.

At 04:51 on the eleventh floor, temperatures of 1,000 degrees centigrade were recorded again on thermal-image cameras. A similar temperature was recorded on the thirteenth floor. Around 06:00 it was believed that approximately 115 people were still unaccounted for.[23] Shortly after, a decision was made not to commit firefighters above the twelfth floor. Within half an hour it was determined that all floors below the twelfth had been cleared of saveable life. Yet at 08:07 the last casualty was rescued from the eleventh floor.

A station manager who was there for much of the incident recounts his experience:

Figure 9: Grenfell Tower, 4:43 a.m. Wednesday, 14th June 2017. (*Wikimedia Commons*)

On the night in question his pager alerted him to a ten-pump fire, and he called the control room to confirm he was en route. By the time he made the call it had already increased to a twenty-pump fire. He travelled through Watford Way, turned right by Cricklewood Lane, through Willesden and then along the Harrow Road to Ladbroke Grove. Radio traffic was heavy and he heard the number of pumps being increased to twenty-five (this message timed at 01:31). As he approached, he looked to his left and saw what he described as a sheet of flame: 'The whole side of the block was alight. It went as far down as I could see and right up to the roof.'

After he parked his car, he walked past crowds of people, many on their phones. Loud crackling and bangs came from the tower and he now had a clearer view of the whole left-hand side, 'totally ablaze from the third to the top.' His initial assumption had been that it was scaffolding alight but he could now see this was far more serious. A 135 ladder was pitched on the east side from the walkway on the second-floor level. Debris littered the ground and he could see large pieces, approximately 75-centimetres-square, blackened and still alight.

He headed for the west side of the building where he had heard over the radio that help was needed and organised two ground monitors to be set up and directed at that side of the block. It was now around 02:15 and it had become obvious that this was something outside his experience. He was aware of cladding fires, but even at this stage he still hoped the fire might remain external. He watched as the fire at the top of the building moved around the roof and started coming back down the tower on the other side:

> I'd never seen that before, a fire coming back down a building. From where I was positioned, I could not see the east and south sides, only the north and west. When it was at the roof I saw it coming across and I remember thinking that the whole thing was going to go up and burn off until we could get water on it.

He left a Watch Manager in charge of the west side and headed for the command unit. It was at this point that he heard a massive 'thud' and heard at the same time someone say, 'he's come out!' He turned to see a police officer and a firefighter attending to a man on the ground about ten feet from the canopy around the tower. Normally, in such a situation the person would not be moved but with debris falling all around it became necessary. To his surprise the police officer found a pulse and they decided to carry the casualty to the south side where we, the London Ambulance Service, were setting up a casualty handling area. This was difficult as he was a large man.

After this he entered the lobby of the tower for the first time where Assistant Commissioner Andy Roe was present and delivering a short speech along the lines of: 'this is a once in a life time incident, we've never faced anything like this before and it goes outside Brigade policy. Do your best.' It was well received, especially among the younger members. It was shortly after this period that he was directed to assist two watch managers with fire-survival guidance calls in the lobby of the block. As he approached, he saw the body of another person on the floor who had apparently jumped. The body was missing a leg.

Debris was raining down and he had to dash the twenty feet to the main doors as a watcher shouted 'clear!' In the ground-floor lobby he teamed up with two watch managers. They told him that the bridgehead had been brought down from the third floor because of the smoke. It was only at this point that he came to realise fire was in fact affecting all the flats.

A system was in place whereby they would receive fire-survival guidance calls from the Command Unit by radio and this information was then written down on paper and handed to crews. They were also recorded on the wall of the lobby area. Normally this would be communicated to the incident commander, but there were just too many. A quick glance at this point revealed thirty to forty calls written on the wall. The crews would then go through entry control with a job in hand already rather than be given an objective at that point as would normally happen. To keep things updated they needed crews to tell them after they came out if they had reached their objectives or not. Thus, the crews themselves, or a runner, would relay the information back. As the bridgehead was positioned on the same floor around a glass door this was fairly easy to do.

The system seemed to be working reasonably well although some returning crews were so disorientated or distressed that they were unable to talk. Others carried casualties straight through to the ambulance crews. This made some debriefs difficult. Residents, too, were often unable to say where they'd come from. Some may have self-evacuated earlier, and others were brought out overcome or simply unable to communicate either through being distressed or not speaking English.

It was around this time that a significant miscommunication occurred. The crews had been given specific tasks related to fire-survival guidance calls. However, as they got to the bridgehead their tasks were changed to firefighting. The reason for this was that conditions on the stairs were such that a decision had been made not to send crews above the tenth floor. The problem was that this hadn't been communicated back to those working on the wall.

> I remember feeling subdued when we discovered crews could not get past the eleventh floor. By this time the Command Unit had stopped receiving anymore fire-survival guidance calls and I think I knew what that meant. Some of us on the ground floor hadn't realised just how bad conditions were on some of the levels above. The eleventh floor had been an inferno and everything beyond that burnt out.
>
> Around 8 a.m. I went outside for the first time in several hours and looked back. Every floor above the eleventh or twelfth was still alight. Gaping holes where windows had been allowed one to see all the way into the building. The concrete seemed to glow pink from the high

temperatures. I realised I was exhausted emotionally and physically. I went back inside and up to the third floor to see the layout and conditions. Hose was everywhere and the stairs were like a river. It was very difficult to navigate them even in fire boots. I realised how difficult it would have been for vulnerable casualties to attempt that journey.

Around 9.30 am I was relieved and collected my nominal-roll board from the command unit. As I walked past [the] casualty-clearing area I could see a body lying, covered.

As London woke to the horror of what had happened hard questions were already being asked.

Table 7: Summary of timeline of fire.

Time	Comments
00:54	Time of first call to brigade.
00:59	Arrival of first fire engines.
01:07	First breathing-apparatus crew enters flat 16 on fourth floor.
01:09	Fire spreads to exterior cladding.
01:14	BA crew enters kitchen for first time. First image of significant fire in external cladding.
01:26	Compartmentation effectively failed.
01:30	Fire reaches top floor of east face, begins spreading laterally on east and north face.
01:42	North face ignited
02:00	Several flats on east and north face alight.
02:06	Major incident declared.
02:25	South face ignited.
02:30	Majority of flats on upper floors of east and north face alight, fire spreads to flats on south face.
02:35	Control decides to revoke stay-put and begins telling residents to get out if they can.
02:47	Stay-put officially revoked.
02:51	Fire spreads to west face.
04:00-04:30	Majority of flats on all floors above fourth on fire or damaged by fire.
08:07	Last survivor rescued from tower.

Media and conspiracy theories

In the days after the fire social media was awash with conspiracy theories and misinformation. I found myself arguing online with strangers. One had seen a tent with stacks of body bags, not realising that such items are routine at a major incident, especially when a significant number of fatalities are expected. Another had seen a number of dark-coloured vans parked and saw this as evidence of something nefarious. Again, it's very normal for those who have died to be transported away from the incident. What colour vans does one expect? The closing of the nearby railway line was viewed by one as suspect, but in reality, with so much smoke, it was likely a precautionary measure taken a few hours into the incident.

Even more bizarre were two mutually exclusive conspiracy theories that carried on for months. One side claimed it was a 'false flag' event, all the survivors and witnesses were 'actors' and the building had in fact been empty at the time. Obviously, it follows that all the hundreds of firefighters, police and ambulance staff were also in on it. You will find people in fringe corners of the internet who believe direct-energy weapons are responsible for a whole host of fires worldwide, including Grenfell.

Far more common as conspiracy theories go were the claims that the death toll was in fact much higher. It was claimed hundreds had died. Television crews interviewed local residents and highlighted these claims. One or two celebrities apparently were expert enough to warrant a slot.

What was forgotten, or conveniently ignored, was the fact the fire brigade command unit had kept a running total of numbers they believed were involved through the night. In the early hours it had stood at approximately 150. This had slowly been reduced as people were rescued or self-evacuated. It's not always easy to get an accurate number, even in a relatively small fire. As the sun rose on 14th June, the figure they had was very close to the final death toll.

In an interview the same morning Scotland Yard Commander Stuart Cundy stated that searching the gutted tower block might take months, but he hoped that the death toll would not run into 'triple figures'.[24] He also stated that they had taken over 5,000 calls to the helpline and, at one point, over 400 people had been reported missing; however, many had been reported several times. One individual was reported missing forty-six times[25]. Here then is where the figure of 400 may have originated.

A little logic would have gone a long way. Out of 120 flats, eighty were two-bedroomed and forty had one bedroom. That equated to 200 bedrooms. With no knowledge at all of the occupants, we could quickly estimate between one or two people per bedroom. Thus 200 to 400 people. The idea that the death toll was over 400 when the hospitals and civic centres were full of injured and survivors should have been seen as an obvious exaggeration.

Still the press picked at the scab. The theory took hold and soon was taken as fact by a significant number of people. A Labour MP joined in only to be taken to task by Andrew Neil in a BBC interview.[26] The fact is that there was one very simple question that could have been asked by Mr Neil or any of the media interviewing people after the fire. It wasn't why they believed hundreds of emergency service workers were involved in some vast cover-up. Rather it was this: 'Of these additional hundreds of people you claim died in the fire can you give a name and flat number?'

By the morning of the fire the emergency services were already trying to compile a list in liaison with the local council. No doubt at some point they had a rough list of flat numbers with residents' names alongside. All those people convinced that there were hundreds of fatalities just needed to contact the relevant authorities with names and addresses.

Some commentators seemed surprised that the fire brigade did not have information concerning who lived in a property. Even if we did, we would have no way of knowing if they were home or if people were staying over. How would a large high-rise block monitor numbers? Do people really expect some invasive clocking-in and clocking-out system? The level of social controls required to maintain an accurate up-to-date record of who is in a residential building is high. Far different at a commercial establishment where people sign in and out.

None of this stopped tongues from wagging. The fact that it took years of painstaking work to finally identify one victim from the King's Cross fire should demonstrate how seriously these matters are taken. A few days after the fire a colleague was sent back to the tower on a relief (the brigade maintained a presence for some days). He found himself with his crew on the upper floors of the tower watching a forensic scientist on her hands and knees sift through the ash. They ended up assisting a little, holding equipment and moving items, and so got talking. They asked what she was doing.

Next to her was a small tray with tiny fragments of bone collected over two hours. They were one part of a small child's remains. This one person would spend many hours painstakingly trying to piece together a poor soul's arm or jawbone to give relatives some peace of mind. The police had also brought in forensic archaeologists. Their work lasted until the December. The statement by the lead forensic archaeologist may be seen on the inquiry website.[27]

While my colleague was watching this expert carefully sift through huge quantities of ash, people online were confident enough to declare that it was all a conspiracy. Ignorant of the work actually being done, they were convinced that a huge cover-up was underway. The press could have helped by better informing the public. However, I saw no attempt at explaining the fact that the emergency services already had a rough idea of numbers a few hours into the incident and that the numbers of survivors and missing in the day that followed tallied up with the numbers believed to have been in the building.

The local council and government understandably received the brunt of the criticism in the aftermath. Yet I could not help but feel disquiet when Prime Minister Theresa May received such aggressive treatment. Ironically, I felt that if we laid all the governments of the last thirty years side by side, hers was the least to blame, having only won a general election the week before.

I understand that the Labour leader, Jeremy Corbyn, also attended on the same day as the prime minister. This was Thursday, over twenty-four hours after the fire. This was the day I had come on duty and been sent a relief from Addington, arriving around midday and staying until the early evening. The only politician I recall seeing was the Labour Mayor Sadiq Khan, walking around with the LFB Commissioner, Dany Cotton.

If I was being generous, I can view such a visit as a genuine show of support. However, what I definitely don't like is using a tragedy as a political opportunity. Truth matters. The truth here is that both major parties have presided over governments under whose watch standards have been eroded. I was therefore less than impressed to read the Labour shadow chancellor, John McDonnell, describe the incident as 'social murder'. Not when several warning fires, Lakanal being the most prominent, occurred under a Labour government.

Whilst Labour made hay with a tragedy they had been brewing throughout their time in power, the Conservatives seemed to be reacting like a rabbit in headlights, unsure of what the best response was. They seemed to be torn between saying that the panels were non-compliant or admitting that there was a problem within building regulations. I got the impression that they would quite like to pin it on one contractor or building inspector or, as a last resort, throw their own council under the bus. Anything to avoid admitting there was a systemic problem within building regulations.

Amid all this mudslinging I don't think I heard one politician mention Approved Document B or The Regulatory Reform (Fire Safety) Order 2005. Nor did I hear anything about the possible relevance of the privatisation of BRE or testing regimes. Several gave interviews where it was quite obvious that they had no idea about fire safety in tower blocks, didn't understand compartmentation or the stay-put policy and had little idea how firefighters fight fires in high-rise buildings. The press in general were no better.

For what seemed like weeks, the focus seemed to be on the fire brigade response and the actions of firefighters. Few seemed interested in explaining the questions about testing and certification, let alone the complexities of building regulations. Those who looked back at the Lakanal fire highlighted the recommendation to consider retro-fitting sprinklers but ignored the far more important matter of Approved Document B being largely unfit for purpose.

Sections of the media, uninterested in facts and nuance, painted a simple picture. Firefighters had told people to remain in a fire. Those who had never heard of stay-put or didn't understand it were understandably horrified. All this had an effect. On fire safety visits we started getting people in domestic houses ask if they were actually supposed to stay put and we had to assure them it was still 'get out and stay out'.

More worrying were the comments from residents in high-rise flats. A significant number of people were unaware of their block's policy at all. Some who were aware believed the advice was now to ignore stay-put. Many times, we had to re-assure people that the advice was still to 'stay put', emphasising the important often forgotten caveat of 'unless you are affected by fire or smoke'. Others were understandably adamant that they would leave in the event of a fire in another flat.

What would have helped in this mass confusion among the public was a statement by a government minister. No such statement came. The press seemed utterly incapable of doing anything but sensationalising and creating false dichotomies.

So, for months, the press focused on perceived failures of the fire brigade, blaming it largely on the stay-put policy with little mention of those who were rescued because of stay-put or those who had died because they attempted to get out. Everyone saw the panels alight, but the insulation attached to the wall behind those panels got much less attention. The lack of fire-stopping, construction of the windows and failure of fire doors got next to nothing. Those media outlets and politicians who bothered to look beyond the first superficial layer banged on about sprinklers but had nothing to say about building regulations.

In short, in my opinion the press and media coverage of Grenfell was mostly poor and often misleading. None of the politicians I heard make comments seemed to grasp the underlying facts.

To combat some of this wall of misinformation I wrote an article that was printed in the *Independent* on 6 November 2019.[28] Despite making it clear that I personally had not attended the fire on the night in question, they went with the headline: 'I was a firefighter at Grenfell Tower. Here's the truth behind the myths'. Not a phrase I agreed with at all. Nevertheless, I was quite pleased to have had the opportunity to make some points clear. In the years since Grenfell I recall only seeing one article written by an expert in building regulations.

Grenfell Inquiry Phase One report

The Phase One report was published in October 2019 with a number of recommendations.[29]

> Building owners should not rely totally on 'stay put' but develop a plan B. Evacuation plans should be developed in case compartmentation failed. Buildings should be fitted with manual alarms for an evacuation signal to be given to all or part of the building.
>
> Disabled residents should be offered Personal Emergency Evacuation Plans.

It found 'compelling evidence that the external walls of the building failed to comply with … Building Regulations'. Instead of adequately resisting the spread of fire, they 'actively promoted it'. The Brigade's planning and preparation came in for much criticism, called at one point 'gravely inadequate'. Incident commanders 'had received no training in how to recognise the need for an evacuation or how to organise one'. Grenfell

Tower had no contingency plan for evacuation. In addition, the entry on the operational risk database was lacking, inadequate, in parts out of date and in other respects inaccurate.

On the incident ground officers failed 'to conceive of the possibility of a general failure of compartmentation or of a need for mass evacuation'. Crucially, it highlighted the time between 01.30 and 01.50 when a decision to evacuate would be likely to have resulted in fewer fatalities. The best part of an hour was lost before AC Roe revoked the 'stay put' advice. Training for an evacuation 'fell by the wayside' despite the Brigade telling its internal monitoring body that work to develop this training was complete by Autumn of 2013.[30] I can certainly confirm that not only did I never receive such training, but I never heard of its existence.

Additionally, communication concerning the numbers and source of fire-survival guidance calls, information about the internal spread of the fire and the results of rescue operations were all poor. Serious deficiencies in command and control on the fire ground was compounded by 'shortcomings in practice, policy and training' in the control room.

It is not my intention to argue about the findings of the inquiry. However, I will make an observation. Firstly, my own analysis prior to the phase-one report was as follows: whilst the first photographic evidence of the cladding being alight was timed at 01:14 it did not necessarily follow that it would get out of control. It wasn't until 01:16 that we see it beginning to accelerate up the building. Certainly, by about 01:19 it appeared to be beyond control.

Barbara Lane stated that 'it is clear to me that the window from 00:58 to 01:40 was when the total evacuation of Grenfell Tower needed to occur'.[31] We saw previously that compartmentation had broken down irretrievably by 01:26. In addition, many residents and firefighters reported smoke in lobbies and the stairway much earlier than 01:50.

What this suggests to me is that any 'window of opportunity' was a little earlier and narrower than 01:30 to 01:50: perhaps 01:26 to 01:40. It was acknowledged that there were difficulties in communicating with residents in a residential high rise.[32] There is no automatic or manual means of raising alarms and no voice alarm announcements such as a Tannoy. Firefighters are thus limited to loudhailers, 999 calls (including fire-survival guidance calls) or firefighters knocking on doors. It was suggested that serious consideration should be given to changing the current approach, especially regarding those who require assistance.

However, changing the current approach is not without risks. It appears that smoke and heat levels within the lobbies and, more importantly, the stairway, fluctuated during the incident. The phrase the inquiry used was 'a decision to evacuate would be likely to have resulted in fewer fatalities'. It would be interesting to see this backed up a computer simulation. We could see how many times out of a hundred an evacuation at say 01:30 could have been successful. We could repeat it for different times. Many of us might find this useful because even now I have no idea what the result would have been of revoking 'stay put' at, for example, 01:40 when multiple crews reported heavy smoke on escape routes.

It is one thing if an evacuation at a specific time between 01:20 and 02:47 (when stay put was officially revoked) has a 99 per cent chance of being successful but quite another if 'likely' just means a 60 per cent chance. A 40 per cent chance of multiple casualties on the stairway is not to be taken lightly. And, of course, those probabilities did not stay constant. They shifted every time someone opened a door, or a door failed. So many variables will be unknown during a live incident.

During the inquiry the commissioner Dany Cotton described the failure to prepare for an incident like Grenfell as like failing to plan for 'a space shuttle landing on the Shard'.[33] She inevitably drew a lot of criticism. At the time many firefighters felt some sympathy. I interpreted the comment as referring to several things failing at once: the windows, the cladding, cavity barriers, fire doors and the lobby smoke-ventilation system. Others, including the inquiry, took a dim view. The media were scathing.

For me the main criticism lay in planning and preparation. Before I researched this book, I regarded the Lakanal House fire as a pivotal point. The phrase 'consider evacuation' in the high-rise procedure operational note seemed to me at the time, and since, as a fig-leaf, almost an 'arse-covering' exercise: 'look we told them to consider it!' Now we know that in fact cladding fires had occurred since 1991 at least. The fires at Garnock Court in 1999 and Salford, Manchester in 2005 should have resulted in a dedicated training package as well as the dangers of cladding being emphasised to trainees and firefighters every bit as much as cylinders and modern roof construction.

One could be generous and point out that the brigade did highlight their concerns with ministers. As late as April 2016 the Commissioner wrote to government ministers warning of blocks with 'significant compartmental issues'. The specific risks from cladding were passed on to local authorities, including RBKC and the TMO whose risk assessor replied that the cladding complied with building regulations.[34]

Yet that does not explain why the PowerPoint 'Tall Building Facades' dated the year before Grenfell was not distributed to stations. Nor why the fire at Shepherd's Court on 19 August 2016 did not prompt a co-ordinated rapid response in regards to training.

Racism

One concern that raised its head concerning Grenfell was racism. Peter Apps writes that, on the night of the fire, 85 per cent of those who died were non-white. He goes on to ask of the council: 'Would it have allowed the kind of conditions endured by the residents of Grenfell Tower to exist in a tower of white residents?'[35] To answer this, one would have to look at the demographics in the other cladding fires mentioned – Merseyside, Glasgow and Manchester. In terms of local authority and governmental response one would need to look at the demographics of other similar disasters over the decades.

My answer would be that the effects of The Regulatory Reform (Fire Safety) Order 2005 affected blocks across the country. The poor standards of maintenance and fire safety are present, regardless of the ethnic make-up of people living there. The most one can say is that blocks with high service charges might spend more money on maintenance. There could be a link with socio-economic class and that may correlate with demographics. But, if that is the case, why not focus on class rather than race?

In any case the questions involving testing, certification and building regulations are far more serious than those of maintenance and general fire safety. Grenfell did not occur because of poor maintenance, bad as that is. It occurred because of the panels, fire stopping, windows and fire doors not complying with regulations. The main driver of the fire spread was of course the panels, and those are on thousands of buildings across the country. In addition, poor building regulations affect every building being built and refurbished, not just buildings with people from one demographic.

In addition, Grenfell did not house people from one demographic. White people were among the various ethnicities. This did not prevent some people pointing to the fire brigade. Yet many firefighters who attended were black or brown. It should not have to be said but everyone is the same colour in a fire. Thick black smoke does not discriminate. Nor does fire. A firefighter cannot see the colour of who they are rescuing. Nor is there a different high-rise procedure depending on the demographic make-up of the block.

Those on the far right made disgraceful comments about immigrants and asylum seekers. Some media outlets made dog-whistle comments either to create controversy or play to the worst instincts of some of their readers. But some on the left were just as bad. Activists and some politicians with no knowledge of building regulations or fire safety or any of the problems that caused Grenfell tried to shoehorn accusations of racism into the discussion. Article after article appeared without a single reference to standards of testing and certification, the BRE, The Regulatory Reform (Fire Safety) Order 2005 or Approved Document B.

Yet it was not racism that introduced ill-defined unqualified fire-risk assessors in the 2005 fire safety regulatory reform order. Racism did not privatise BRE in 1997. Nor was it racism that allowed products to slip through testing, obtain certificates and be marketed for high-rise buildings in the 2000s when, instead of adequately resisting the spread of fire, the product actively promoted it.

A view from a fire safety officer

An experienced fire safety officer attended Grenfell later on the day of the fire. He described a harrowing image:[36]

> The top of it matt black with thick soot. Most of the windows of the twelve or so floors I could see above the trees were emitting rolling dark grey smoke, occasionally punctuated by a flame here and there. The few windows that did not have smoke coming from them were even darker blackened holes.

He has kindly let me quote directly from his blog, *Stevedude68: The life of a London Firefighter*.[37]

Stay-put is described as 'a decades old strategy, incorporated into the design of a building, primarily for purpose built residential buildings, whereby the building is generally made of concrete and each flat is a concrete box able to contain a fire until it burns itself out'. Thus, in theory, even if the Brigade failed to arrive, so long as doors and windows were shut, the fire should burn itself out, being starved of oxygen or fuel.

At Grenfell we now know that the cladding was not only non-fire resisting, it was actually flammable: 'Like covering the building with petrol'. His experience of high-rise fires mirrors mine in that in scores of high-rise fires most were confined to a room or two in a flat with a small number spreading externally a floor above. He describes five decades of UK firefighters dealing with fires in high rise blocks without incidents like Grenfell.

The stay-put policy is 'a strategy related to the building and agreed as part of the building's overall firefighting strategy at the design stage. It has nothing to do with the geography of the area'. The provision of one staircase and one or two lifts relies on the fact that each individual flat is in effect a separate fireproof box.

Office blocks, hotels, cinemas and other places with large numbers of the public are often evacuated fully. They tend to be open plan, more prone to fire spread and have significant numbers of people in an unfamiliar layout. Importantly:

> Firefighters attending a fire are not the ones who decide whether or not it is a 'stay put' policy. Blaming Firefighters for a failure of a stay put policy is like blaming a mechanic who can't fix the engine of a badly built and poorly maintained car after the engine has seized up.

He goes on to state, 'That type of fire spread had never been seen before, it was completely beyond the knowledge, experience and frame of reference of anyone in the UK Fire Service.' Whilst the inquiry suggested an early evacuation would *likely* have saved lives this senior officer with thirty years' experience states, 'I have no way of knowing how the building could have been instantaneously evacuated at that point'.

> if a hundred or more people had opened their doors and evacuated simultaneously, as they would in a hotel for example when the alarm sounds for a fire which is still small and picked up quickly by the fire detectors what would the outcome have been? ... An understanding of human behaviour in fire as well as knowledge and long bitter experience of fire, heat and smoke leads me to conclude that my colleagues and the London Fire Brigade would now be facing accusations of sending

a hundred-plus people to their deaths. Instant blinding choking smoke, heat which saps strength in seconds, blind panic.

He describes vividly the reality of being in a fire compared to peoples' perceptions. No running through 'smokeless flames' with a wet handkerchief covering the mouth. Instead, blackness. Unable to see your hand in front of your face. Searing heat would drive you to your hands and knees and then your belly. The acrid thick smoke would choke immediately.

The phase one report was published in October 2019. Some disquiet was expressed at the focus of the report. On the subject of a window of opportunity, he had this to say:

Even if by some miracle of divine intervention that decision was made at 01:30 to give them the best chance, how was that to be communicated to the hundreds of residents still in the tower at that time?

There was (for very good reason that I will not go into here) no public fire alarm. In the noise and confusion would people, especially those who were still blissfully unaware at that time, have heard or taken notice of loud hailers. Evidence from some survivors and tragic testimony from some of those who were trapped clearly demonstrates some people did try to escape and, facing choking blinding smoke & fumes, either went back, went further up into the building to escape the poisoning atmosphere or got no further than opening their front doors.

It is true that a number of people took that brave decision and were able to escape, some barely conscious as they got to safety, others collapsed and rescued by Firefighters on the stairs. But if the order to evacuate had been communicated and heard, how were the LFB meant to encourage those people to make an orderly escape?

Witness testimony from many Firefighters for those who have been bothered to read it, is littered with reports of crews who did reach people in flats on upper floors where they often refused to open doors or found escape untenable and remained in their flats, or, most tragically Firefighters, physically exhausted from the climb to those higher floors in heat, smoke and debris, quickly realised that to remove people was to condemn them to a certain death within minutes of leaving, at that time, a relatively safe environment, not being able to comprehend the spread of fire that was to follow.

I am afraid, for all of the great minds and detailed analysis and investigation into the night of the fire, this seems to have been overlooked. I admit to having had no sight of the report as yet, but even if the facts I mention above have been included, they have been overlooked in what appear to be conclusions built solely on technical analysis without any consideration of human behaviour, emotions, lack of experience of this type of failure anywhere previously and an understanding of the utter horror those responding, trapped or witnessing had to endure.

In summary, I can only conclude with as much objectivity as I can muster having been involved in the incident, that although mistakes were made by LFB, these were not reasonably predictable in terms of the rapid deterioration of events on the night and as such the conclusions of the report have, in my opinion, almost been pre-determined to scapegoat the London Fire Brigade and its personnel to what end?

Systemic Failure: Those are the headlines today in relation to the LFB. I'd argue that systemic failure has appeared everywhere in the sorry tale of the Grenfell Tower Fire. From de-regulation of Fire Safety laws in the early 2000s, from the apparent cost-cutting and poor oversight of the refurbishment of the Tower, to the way in which the inquiry was set up about face and the conclusions drawn at the end of phase one.[38]

His recommendations point to the root causes of this tragedy: 'regulations around construction, refurbishment, materials used and fire safety legislation will also change. It needs to if we are to prevent another tragedy in the future'. This, then, is the opinion of a highly experienced fire safety officer, the sort of person newspapers and media outlets should have been interviewing.

Sprinklers hats and parachutes

We began this book with the analogy of the airline industry and the dumbing down of standards from the manufacture of a simple bolt to all the testing, maintenance and inspections one hopes are standard. Let us imagine a second scenario, one where standards are relaxed to allow flammable clothing. There have indeed been examples of quite serious injuries, some involving children's dresses and costumes catching light from candles. Thankfully, there are regulations covering such things (for example The Nightwear (Safety) Regulations 1985). But let us suppose this was all quietly ditched and manufacturers were not just allowed but encouraged to manufacture and supply flammable clothing.

Fire safety officers and experts from the industry might raise concerns. The media would have little interest in such boring and dry subjects as clothing regulations and fire safety. However, after some serious incidents, and perhaps a particularly newsworthy one involving multiple casualties, the press opens a lazy eye. Who does it turn to for answers. Fire Safety Officers? Experts responsible for writing the original regulations? Do they allow time to investigate the complex nuances of the subject? Of course not. It turns to politicians and the odd celebrity for a two-minute interview from which it garners the odd soundbite.

Then let us suppose the press latches on to a possible solution: sprinkler hats! A very easy concept to understand. Bound to save lives. Everyone can carry little water extinguishers connected by tubes to headgear. Far easier for the sitting government than re-writing all the relevant regulations and getting rid of all the flammable clothing. Or admitting they were part of the problem in dumbing down the regulations. Then let us imagine a world where the media and politicians devote far more time speaking about sprinkler hats than the fact that we've dumbed down the regulations to such an extent that we've allowed a situation where loads of people are wearing highly flammable clothing.

Of course, in this hypothetical world experts and firefighters would be screaming at their televisions every time the subject came up: 'Just stop wearing flammable clothing!' But no one would bother interviewing them because sprinkler hats are easy to understand whereas re-writing specific regulations about the flammability of clothing doesn't make good TV. This, of course, is a ridiculous scenario. Yet that is precisely what we have been doing when discussing buildings. Everyone prioritising sprinklers over building regulations and practices.

Don't get me wrong. Sprinklers in buildings are a very good idea. Very effective in reducing injuries and fatalities. But just hear me out for a second. Firstly, ask yourself why is it that we've successfully dealt with fires in high-rise residential blocks for decades without any sprinklers? By all means fit them. I will be first to support this.

Of course, we'd have to ensure that those fitting them understand the importance of compartmentation. Otherwise, there will just be many more buildings with potentially failed compartmentation. One expert recalled how one residential high-rise 'had been converted into a sixteen-storey chimney by someone with a Kango Hammer just punching a series of holes through concrete slabs'.[39] Retro-fitting sprinklers in old blocks before we overhaul building regulations and control would likely create many more potential Lakanal House or Grenfell Tower type situations.

So, before we think about fitting a single-sprinkler head what I'd like to see first is the relevant regulations re-written. This should be the priority. It likely involves, among others, Approved Document B and The Regulatory Reform (Fire Safety) Order 2005. At the same time, make guidance notes clear. It doesn't take weeks (let alone years) to write a regulation stating clearly that clothing should not go up like a rocket when it comes into contact with an ignition source.

Nor should it take long to make a similar statement about the exterior of buildings. The next priority should be to remove the risk. To be fair, removing thousands of panels from hundreds of high-rise buildings is not an easy task. But this should take weeks, not years. There is no excuse for allowing a building with such potentially hazardous materials to be occupied.

I can agree with some of Peter Apps' 'wish list' regarding new builds.[40] A second staircase in new buildings would be beneficial in terms of firefighting and evacuating buildings. An alarm to warn residents to evacuate would most certainly have helped at Grenfell. Sprinklers would significantly assist in fire suppression. But the priority should be the underlying cause and that means overhauling the system from manufacturers to enforcement but specifically building regulations and control.

Stay put

Let us now turn to the future of the stay-put policy. Readers may be surprised to learn it remains the policy in many high-rise blocks across the country. In its defence, between 2009 and 2010 in 8,000 fires in blocks of flats only twenty-two required the evacuation of more than five people.[41] It followed that blocks generally do not need a communal alarm. One expert, described as 'a vociferous advocate of stay put', advocates that it is advantageous to disabled people for a block to be built in a way that evacuation is not required.[42] Compartmentation is supposed to be robust.

Perhaps more importantly at Grenfell eight people were recovered from the lobbies and stairs and it is believed that they attempted to escape but were unable due to the conditions.[43] We should not try to fight the last war. The next fire might well have a higher number of fatalities on the stairs.

However, another witness described the UK as a global outlier in its reliance on stay put.[44] Peter Apps describes a 'misguided fear of panicked evacuation, which is particularly prevalent within the fire service'.[45] He describes it as a 'convenient policy' which prevents blocks from requiring a second staircase. Also, it would require alarms or someone to manage evacuation. If we accept the probability of fire breaking out of a compartment, the argument for sprinklers becomes more persuasive. He goes on to say, 'acting as the enablers of an evacuation does not fit with the way the British Fire Service sees its role.'[46]

But is fear of a panicked evacuation truly 'misguided'? And should first responders be responsible for evacuation in one type of building when, in every other type, evacuation is carried out by trained staff before the brigade arrives? Allowing them to focus on dealing with the incident.

My argument would be to ask why the policy worked for many decades. It worked when building regulations, building control and enforcement were robust. The fact that we have spent thirty years merrily dumbing down these things and forgetting why compartmentation is there is not a reason on its own to abandon stay-put. My first answer would be to build, refurbish and maintain buildings to the standard we enjoyed before we covered buildings in solidified petrol and forgot the importance of compartmentation.

However, I can see that a policy may no longer fit for purpose once people have lost all faith in it. Even though it was never simply 'stay-put' but rather 'stay-put unless you are affected by fire or smoke in which case get out if you can'. What then is the alternative?

I would argue that the worst option is to rely on firefighters to decide when to change the policy. They are not there when the fire starts. They might arrive five minutes later. However, in busy periods, or whilst a large-scale incident or two are in progress elsewhere, they might not arrive for twenty minutes. It seems rather foolish to have a policy that relies on a decision being made by someone who is not there in the initial stages and for whom there is no guarantee when exactly they will arrive.

Even then we are asking someone to weigh up the risks of stay-put against the risks of placing significant numbers of people on the, often, only stairway. They have to be confident that the exit route is safe and *will remain safe for the duration of the evacuation*. Yet there is no way of knowing all the variables. It is debatable, despite the identification with hindsight of a 'window of opportunity', if anyone could have reasonably known that at Grenfell. It is at least possible, if not likely, that at some point there will be a mass casualty event on a stairway in one of the many thousands of high-rise incidents each year. In addition, the fire brigade does not arrive with multiple spare people to form 'evacuation teams'.

Should they go down this route, some way of communicating with residents would have to be provided, a tannoy system or an alarm accessed via a drop-key. It would not be enough to leave it to control operators to tell only those who ring 999. The government would have to accept that the fire brigade is no longer just a fire and rescue service. It would have to become a fire, rescue and evacuation service. In that case the number of firefighters would have to be increased by a significant factor.

In addition to that, many brigade policies would have to be re-written. Many of those have been imposed after inquests and coroners' recommendations following fatal fires. It is difficult to see how an evacuation could be achieved quickly if every crew committed had to be under air and armed with a 45mm jet, let alone the fact that they are not supposed to go past or above a fire. We will come back to this shortly.

Thus, my first option would be a twenty-four-hour concierge. Concierges could manage evacuation before the brigade arrives, as in commercial buildings and schools. They would also help improve the general maintenance and fire safety of the block. Defects in fire lifts and smoke-ventilation systems would be dealt with quickly. They would also be on hand to operate any such facilities for the brigade. It is very unlikely that an initial crew would be familiar with every type of smoke-ventilation system. A concierge would also be familiar with who was in the building and assist with the evacuation of disabled people, or inform the incident commander of the details.

The downside is this would be expensive. Perhaps four people on a rolling shift might mean over £100,000 spread across the flats. Workable on large blocks, but those with perhaps just thirty flats would have a bill of over £3,000 a year each.

The other alternative is to fit a communal fire alarm, perhaps connecting this with smoke alarms in individual flats. An alarm simply in communal areas would only sound when fire had breached the initial compartment. An alarm connected to alarms in flats in some blocks would have everyone evacuating. The latter would be happening on a regular basis. Some blocks we attended once a year or less but others we had a call every week, sometimes more. There will inevitably be some blocks where a significant proportion of people start to simply ignore repeated alarms.

Of course, there is not supposed to be any fire loading in communal areas of high-rise blocks. In theory, alarms in communal areas would actuate only if compartmentation was breached, or someone opened a front door allowing smoke into the lobby.

The problem with integrated alarms in flats concerns access, whether for fitting, maintenance, testing or resetting. New laws extending powers of entry would likely be resisted by some. Whatever decision is made it seems rather to miss the point if compartmentation is continually compromised from poor building regulations, building control, fire safety and enforcement.

A related factor concerns those who may have difficulty evacuating a building. Guidance published in 2011 stated that measures must be in place to ensure they could leave the building in an emergency, importantly without reliance on firefighters.[47] The use of Personal Emergency Evacuation Plans, PEEPs, is advocated to facilitate this. In contrast, guidance encouraged reliance on stay-put, the thinking being it was 'usually unrealistic … to have in place special arrangements'. Fifteen of the thirty-seven residents who had a disability died at Grenfell.[48]

How would this work in practice? Peter Apps suggests that many disabled residents have people living with them or someone close by who can help out.[49] Others can be rehoused on the ground floor. I feel this does not answer all the questions. What of visitors who are disabled? Do they require a PEEP before they can visit? Are we to ban people with disabilities from certain buildings because they don't have a PEEP? What about a resident who becomes disabled? Will they be forced to move if a PEEP is delayed or find that there is no practical way to assist them within the length of time an escape route is supposed to be protected?

On the plus side it would be useful to firefighters to have a record of residents in a building with a PEEP, either on the operational risk database or within a building plans box at ground-floor level. It would be even more useful to be met by a concierge who could give details as firefighters arrive. However, none of this is a substitute for robust building regulations that ensure compartmentation does not fail.

Customer service and enforcement

Recently, I was stuck in the Kafkaesque awfulness of a telephone automated system trying to get through to customer services. Round and round in circles only to be confronted with someone who could not help at all, undoubtedly living their own small nightmare of working in a call centre. Experiences from both ends of the phone I am sure many are familiar with. It occurred to me that we accept this as normal now and all but a few have forgotten what good customer service is.

It reminds me of what a colleague, Rob Farrant, once said to me about quality. We think we know it when we see it but it's hard to define. Conversely we often know when it is absent. Think of examples of famous paintings or sculptures and then compare them to some examples of modern art. Or some wonderful feat of engineering or sporting prowess. Why is there some consensus as to which football players are world class and which are not? Why are some works of art considered classics? What makes one building beautiful and another an ugly eyesore?

Obviously there is difference of opinion and beauty is in the eye of the beholder. But out towards extremes the difference becomes obvious. A functioning public service against one that fails on the outcomes the public cares about. The problem is, as hard as quality is to define it's even harder to measure. Much easier to measure things that can be counted. It is easy to count the number of fire safety audits, much harder to assess their quality. But without an essence of quality we are left with only things that can be counted but which may not be worth very much.

Let me attempt to define what good quality customer service might look like. I would suggest first a phone number for customer service is easily found on literature or a website. It's not hidden or, worse, non-existent, forcing you to chat with a robot online. Instead, you ring the number and, within a reasonable number of rings (let's say less than ten), a human being answers on the phone. This may come as a shock to younger readers, but this was quite normal once upon a time.

You tell that person your problem and they immediately understand the nature of the problem and are able to help you. They aren't reading from, or confined to, a script. If they can't help, they put you through straightaway to someone who can (again within ten rings someone picks up the phone). In the rare occasions no one is available a promise is made to call you back.

The Fire and Aftermath 145

Later someone calls you back and they are aware of the problem and can rectify it.

This may all seem a pipe dream now. Instead, what we have allowed to evolve is not customer service at all. It is a fig-leaf. Like the man behind the curtain in the *Wizard of Oz* desperately pulling levers to give the illusion of control. In my more cynical moments I suspect it is designed to tempt people to give up contacting customer services. No doubt organisations will claim they have streamlined the system so that they are more efficient at helping people. That isn't how we, the public, experience it.

Which brings me to building regulations and fire safety. It seems to me that we have allowed building control and fire safety enforcement to wither away. Of course, they still exist. The buildings and offices are still there. There's a contact e-mail or phone number. But what does this matter if a team of experienced officers have been whittled down to one? What if the workload is such that one person can't possibly deal with the huge pile of cases adequately? And any powers that might have existed have been chipped away over many years. It is a paper tiger. A hollowed-out shell of a system. Like so many of our public services.

Turning to Grenfell, what would a good response have looked like? First, the leaders of both main parties would have come out together a day or two after the fire. They would have apologised and together acknowledged that they were equally responsible for the dumbing down of standards in testing, maintenance, building control, inspections, fire safety, enforcement and provision of emergency services.

They would have listed the relevant legislation that contributed to the situation. One of the first and most important priorities would be to overhaul building regulations, especially Approved Document B. In the short run it should not be difficult to clarify the requirement not to cover buildings in highly flammable materials.

A plan of action would include testing and removing similar panels in a reasonable time. Three months would be a reasonable time. Six months at the most. The cost would be borne by the taxpayer, unless an investigation found that a contractor or other third party was negligent. It definitely would not be borne by flat owners.

They might well accept that sprinklers should be retro-fitted but they would rightly decide that not a single sprinkler head would be fitted until building regulations and building control were rigorous enough to guarantee that any works did not affect compartmentation. There is little point in starting a huge work programme before standards in building regulations and control are not of a sufficiently high standard.

At the same time The Regulatory Reform (Fire Safety) Order 2005 should be re-written. Fire risk assessors need to be defined more precisely and require a level of training and regulation. Newspaper articles would be written by experts in fire safety and building control. Television companies would interview experts. Programmes such as BBC Question Time and BBC Politics would invite such experts to inform the public of the main elements.

Cause

It should be clear by now that there is not just one cause for the tragedy. One expert witness stated:

> In my opinion therefore, irrespective of all of the non-compliances … at the time of the fire the situation was that the use of the Reynobond 55PE Cassette; CelotexRS 5000; Kingspan Kooltherm; Kingspan TP10; Celotex TB4000; the Aluglaze Styrofoam core insulating panels; and the Siderise Lamatherm RH25G cavity barriers; when considered separately or as a single assembly, were fundamentally noncompliant with the Building Regulations.[50]

This list of failures on the night should be seen as the symptoms, not the cause: flammable panels; combustible insulation; lack of fire stopping; gap around the windows allowing a path for fire spread; flammable filler in places; the failure of, and defective, fire doors; and smoke ventilation system and fire lift not working as they should.

Instead, the cause is far more systemic within the entire chain: manufacturer; testing; certification; building regulations; building control; maintenance; inspections; and enforcement. It has been stated by one expert that 'the fact of the matter is as soon as a component failed … there was almost nothing that could have been done to mitigate the consequences'.[51] The pattern of fire spread and development, and the lives lost, had their 'origins in decisions, acts and omissions that occurred years before, some as far back as three decades'.

A review of building regulations by Dame Judith Hackett concluded there had been 'a race to the bottom' in the construction industry.[52] The accepted official position was that combustible cladding was already banned and thus the matter was one of non-compliance rather than any systemic ambiguity. The government would eventually announce plans to ban combustible materials in buildings over eighteen metres high. Blocks between eleven to eighteen metres could continue to use the products if they passed a large-scale test. Peter Apps wrote, 'With our system of privatised testing, certification and sign-off wholly failing to enforce the rules, combustible insulation was being installed on thousands of tall buildings around the country'.

We recall that Prosser and Taylor labelled the problem 'benign neglect'. They wrote that, once the fire had spread from the fridge to the kitchen to the underside of the panels, 'there was almost nothing that could have been done to mitigate the consequences'.[53] As already noted, those consequences 'had their origins in decisions, acts and omissions that had occurred years before, some as far back as three decades'.

The FBU described the Grenfell Tower fire as a: 'a crime caused by decades of deregulation, privatisation and the prioritisation of profit over safety'.[54] Whatever phrase one uses, it should be clear now that the multiple failures at Grenfell, and at similar fires over the years, were caused by the widescale, systemic dumbing down of standards across the board.

Having identified the main underlying cause, who then to blame? Unsurprising that survivors and relatives branded the companies involved as 'little more than crooks and killers'.[55] But this is just part of the picture, one layer of a multi-layered problem.

The blame game

The various parties involved in the debacle of Grenfell have spent much of the time trying to deflect blame onto each other. There are many layers to this. We could make a wider political point. The first layer is a hollowing out of public services in general. Those readers familiar with the police, probation, social work, care homes, NHS and a host of other areas will have their own stories. Whether it's when only one police car is available across the largest London Borough, or a building control officer finds himself overwhelmed with cases.

The second layer, an ideological drive towards deregulation, in a seemingly misguided belief that all regulation is bad. Some regulation might well be badly conceived but others, such as wearing a seatbelt, have clear benefits. As does not covering buildings in flammable material.

But let us just focus on Grenfell. If the problem involved only panels, then perhaps we could blame one rogue manufacturer or contractor. Yet it was not just the panels. The way the windows were fitted had a significant impact, as did the failure of fire doors. If one testing certificate was inaccurate, we could blame one test or one organisation. Nor does it affect just one block or one local authority. It affects hundreds across scores of local authorities, of a variety of political persuasions. The fact that scores of different building control inspectors and fire risk assessors have allowed similar panels to be fitted on hundreds of buildings over many decades should demonstrate clearly that the problem is wider and deeper than Grenfell alone.

We also cannot pin it on one piece of legislation or even one particular government. Do we blame Major's government for not responding to the Knowsley Heights fire in 1991 or privatising BRE in 1997? Or perhaps Labour, under whom many of the tests and certification took place? In my opinion the Fire Safety Regulatory Reform Order in 2005 was certainly a contributory factor in the deterioration of standards of maintenance and fire safety within many high-rise blocks. The Lakanal House fire occurred in a Labour Borough under a Labour government.

I would happily lay a fair amount of blame on the Conservative and Liberal Democrat coalition government. They had plenty of opportunities to act on the recommendations after the Lakanal House fire. Worse was the drive for deregulation and inaction on anything pertaining to the known risks of flammable cladding. The subsequent solely Conservative governments doubled down with this appalling approach. Yet that government did not create this problem. They inherited it and made it worse. Does that make them more to blame or less? Ironically Theresa May's government, being just a week old, was perhaps the least to blame.

The sad fact is that they are all to blame. Blaming one particular council, political party, manufacturer or contractor misses the point. If we are only looking for a conviction, or a political scalp, we will miss the bigger picture. This is my biggest fear: missing the wood for the trees.

For example, one of the key moments in our timeline came after the 2005 cladding fire in Manchester. A subsequent amendment to a passage in guidance was restricted to a paragraph headed 'insulation materials/products' stating 'any filler material' should be of limited combustibility.[56] If this guidance had been clearer and had specified that all materials in cladding systems on exterior of buildings should be non-combustible, Grenfell may have been avoided. It is in such details that future tragedies will be prevented.

Conclusions

The Inquiry phase one report listed a number of shortfalls regarding the London Fire Brigade. It was described as 'conservative, cumbersome and resistant to change'.[57] I highlighted the areas of planning and preparation. It is much harder to criticise the firefighters on that fateful night. They were confronted with something that none of their training or experience had prepared them for. Even with hindsight and having trawled through the evidence, the prevailing feeling of those officers who did not attend is one of 'there but the grace of god go I'. If we go back in time to, say 01:40, and make the decision to evacuate I still do not know what the result would be. If experienced firefighters are unable to judge when or if an evacuation could have been carried out safely that alone should give some pause for thought.

A new strategy for evacuating a high-rise has since been proposed. Unsurprisingly, it boils down to sending a crew above the fire, in safe air, possibly wearing breathing-apparatus equipment but not started up and beyond the entry control point, to knock on doors. Equally unsurprisingly, the union argued that it went against a number of procedures and policies.[58]

It would be nice to report that the questions arising from the Grenfell Tower disaster have been resolved. Unfortunately, a disaster has morphed into another scandal. As of 27 December 2023, more than 180 residential buildings in London still had patrols known as 'waking watches' after failing safety inspections.[59]

Government estimates are that between 6,220 and 8,890 medium-rise buildings will need 'life-critical' safety work on top of more than 3,500 high rises.[60] A survey shared with the magazine *Inside Housing* showed that only 21.8 per cent of leaseholders in dangerous blocks had seen remediation work start. For 44.1 per cent a date had not even been identified for work to begin. The majority had no timescale for the works and only around 10 per cent expected them to be completed within the following twelve months. Many leaseholders are still trapped with 78.8 per cent unable to re-mortgage.

As I finished the draft of this book a four-part drama, *Mr Bates vs The Post Office*, aired on ITV. It detailed the horrendous scandal of hundreds of sub-postmasters wrongly prosecuted and convicted for fraud. It gripped the nation partly because many people recognised the Kafkaesque, bureaucratic nightmare of a faceless system refusing to listen. A 'computer says no' situation. The government responded immediately, which was ironic given the fact that governments of all kinds had failed to respond since 1999 when the scandal began. Now we seem to have policy by TV drama or documentary.

No doubt the many thousands caught up in the cladding scandal found this all depressingly ironic. Hoping for themselves that a TV drama about their own plight would have a similar effect. The survivors of Grenfell have had their documentaries but still await some sort of feeling of justice.

I am sure we can be better than this. We don't need to wait for a horse to bolt before we shut a gate. We should not always need a disaster to make things safer. Governments should not require a TV drama to spur them into action. Local authorities don't need a Grenfell to carry out adequate risk assessments and ensure fire safety in their stock of high-rise blocks.

Over 350 years ago the Great Fire of London destroyed much of the city and heralded in a host of new building regulations in the Rebuilding Act of 1667. One of these required all houses or buildings, whether great or small, to be built only in brick or stone. It is from such basic common-sense decisions that later regulations in the last century required the exterior of buildings to resist the spread of fire.

Yet, somehow, we as a society have forgotten lessons of the past and allowed the following situation to develop over many decades: a situation whereby scores of building inspectors interpreted building regulations to allow hundreds of buildings to be covered in flammable materials across scores of councils and under governments and local authorities of all political persuasions.

Appendix

Timeline to Tragedy

1984: Part-privatisation of enforcement of rules: Building Act (1984) introduced Approved Inspectors, independently monitored and regulated by the Construction Industry Council Approved Inspectors Register (CICAIR), to carry out building control work in England and Wales. Existing building regulations (350 pages) replaced by twenty-four headline standards. Rules and guidance concerning fire contained in *Approved Document* B.[1]

1991: Cladding fire in eleven-storey block at Knowsley Heights, Merseyside. Called into question Class 0 test and raised the question of use of flammable cladding. Instead, focus was on inadequate fire barriers. Expert witness to inquiry stated that, if this had been acted on, the 'entire crisis may never have happened' and 'it is impossible to overstate the importance of what we missed here'.[2]

1994: Mock tests at BRE identified problems with Class 0 cladding panels such that they 'suffer extensive surface burning', often spreading nine metres to top of test building. No changes were made to guidance.[3]

1997: BRE privatised, now required commercial income to fund work. Regulations required rain-screen cladding to meet Class 0 but for insulation to be of limited combustibility.

1999: Cladding fire at Garnock Court, Irvine, Scotland. Fire spread from fifth floor to roof.

1999: A subsequent report by Dr Moore, expert from Fire Safety Development Group, to parliamentary committee explained the difference between limited combustibility and Class 0 tests. A material could be combustible and still achieve a Class 0 rating by adding fire-retardant chemicals or facing the combustible material with a metal foil or sheet: 'this serves to undermine the integrity of the regulations'. MPs advised ministers to scrap Class 0 standard and require cladding systems to be entirely non-combustible or pass a large-scale test at BRE.[4]

2000: Government does not follow advice. Instead retains Class 0 standard and introduces large-scale test at BRE as an alternative.

2001: In 2001 at the BRE a test of a cladding system consisting of non-combustible glass wool insulation and ACM cladding panel failed disastrously. Flames extended twenty metres, twice the height of the nine-metre rig, in less than six minutes. This was despite ACM being of Class 0 and supposedly being able to be used on tall buildings in compliance with guidance in Approved Document B. A subsequent report from BRE to government the following year did not make clear the seriousness of the situation.[5]

2002: The CWCT, representing the cladding industry, warned new testing methodology which resulted in a number of fails could result in abandonment of use of rain-screen cladding with economic consequences for the industry.[6]

2004: The Central Fire Brigades Advisory Council was scrapped and national standards abolished. This removed central control over policies.

2004: Arconic, the firm that supplied the rain-screen cladding panels at Grenfell, conducted a test on its ACM panels in France. The panel came in two types, riveted and cassette. The cassette system failed completely, burning ten times as fast with seven times as much heat and three times as much smoke. It failed to achieve even the lowest grade E.[7] Arconic continued to sell it.

2004/5: The now privatised BRE reported to government to consider introduction of smoke toxicity limits for materials used in internal walls and ceilings as in most of Europe. UK did not follow.[8]

2005: EU adopted new standards from A1 to E. UK decided to use Euroclass B but retained Class 0 (equivalent to Euroclass C or D) as an alternative.[9]

2005: Kingspan, the company responsible for much of the insulation used at Grenfell, had its K15 insulation tested at BRE. It was used in combination with heavy cement-fibre cladding panels. BRE stated the test was for this exact system only. Kingspan marketed it as 'successfully tested to BS8414' and 'acceptable for use above eighteen metres'. They did not mention that it was for one specific system only, i.e. with heavy cement-fibre cladding panels. They also marketed it as achieving Class 0 even though it 'was not relevant' and only obtained the pass by ripping off foil normally attached.[10]

2005: Introduction of Fire Safety Regulatory Reform Order which allowed for ill-defined, unqualified fire risk assessors.

2005: Just to emphasise the dangers, in 2005 there was another cladding fire in a nineteen-storey block, The Edge, Salford, Manchester. The panels had a Class 0 rating and had been used in compliance with Approved Document B. A subsequent amendment to the passage in guidance was restricted to a paragraph headed 'insulation materials/products' stating 'any filler material' should be of limited combustibility.[11] If this guidance had been clearer and specified all materials in cladding systems on exterior of buildings should be non-combustible, Grenfell may have been prevented.

2006: Arconic applied to British Board of Agrément (BBA) for a certificate for the ACM panels based on the test where panels were riveted. The cassette type test was omitted although the diagram provided showed both types. They also claimed the panels met Class 0 but only provided a 2003 test on a 'fire retardant' version. The British Board of Agrément (BBA) issued the certificate.[12]

2007: Arconic manager attended industry conference where it was acknowledged that a building covered in ACM would act similar to a 19,000-litre oil truck and 'kill 60-70 persons'.[13]

2007: Kingspan changed the chemical formula in K15 and added perforations to the foil. A second test (after the 2005 test) failed, despite the use of non-combustible solid aluminium cladding panels in front of the new K15 insulation. It became a 'raging inferno', nearly destroying the lab.[14]

2008: Further tests within Kingspan resulted in 'rapid and serious failures' of K15.[15]

2008: Despite test failures, Kingspan approached BBA for certificate, providing only 2005 test with cement panels. Certificate was issued.[16]

2009: Kingspan obtained certificate from LABC, London Authority Building Control, which recorded it as 'material of limited combustibility' suitable for buildings over eighteen metres.[17]

2009: Lakanal House fire involving cladding panels beneath windows. Six people died. Subsequent tests by BRE found the panels 'burned fiercely and could not even obtain the limited Class 0 classification.'[18]

2009: Four years after the introduction of FSRRO (2005), the TMO responsible for Grenfell were relying on in-house staff to carry out fire risk assessments. The LFB requested that they appoint a competent risk assessor or receive an enforcement notice.[19] The TMO procured a firm called Salvus who produced a report raising serious concerns. The TMO would later dispense with them and turn to a consultant who might be more willing to 'challenge the LFB on thorny issues'.

2010: Conservative and Liberal Democrat coalition government comes to power. Following year announced a 'red-tape challenge' which resulted in a resolution to 'kill off health and safety culture for good'. Policy of 'one in, two out' later became 'one in, three out'. This made new regulations related to Approved Document B and fire safety effectively 'impossible'.[20] Subsequent policy of austerity resulted in building control team at RBKC falling from twelve inspectors to five. Across the country in the decade prior to 2021 number of council inspectors fell by 27.4 per cent.[21]

2010: Over next seven years austerity cut fire service personnel by 25 per cent.[22]

2010: Fire in flat on sixth floor in Grenfell Tower prior to refurbishment. Compartmentation prevented fire spread. However, smoke-control system allowed smoke to leak back into building on fifteenth floor lobby. TMO discovered vents did not close properly but delayed repairing it until planned refurbishment years later.[23]

2011: Further test of ACM panels in cassette form failed again and they were formally categorised as E, making them inappropriate for high-rise buildings across much of Europe. Arconic removed B grade from marketing material after 2012 fire in France but did not mention change to grade E for cassette type.[24]

2011: RBKC and KCTMO discuss plans to refurbish Grenfell Tower.

2012: Cladding fires in Paris, France and Tamweel Tower, Dubai.

2012: BBA carried out a review of the 2006 certificate for ACM panels, now reduced from Euroclass B to E. Arconic failed to respond and

	continued to list only fire-resistant product (which was Class B) on their website. The BBA re-issued the certificate.[25]
2012:	Architect Studio E asked to carry out work despite no experience on Grenfell Tower. Studio E selects solid zinc panel for cladding. At a meeting in April with CEP Architectural Facades cheaper ACM panels are suggested. In August consultant e-mails Studio E regarding insulation made of combustible plastic. Neither are aware that it did not meet building guidance standards. Fire safety consultant Exova in the fire strategy for the refurbishment plans makes no mention of plans to clad the building and says that the work is expected to have 'no adverse effect on the building in relation to external fire spread, but this will be confirmed by an analysis in a future issue of this report'.
2013:	Further tests on ACM panels confirmed the E grade classification of the cassette system and moved the riveted system from B to C. An Arconic employee testified to inquiry that she did not tell clients in UK as she did not believe European standards were relevant to the UK.[26]
2013:	Inquest into deaths at Lakanal House fire recommended Approved Document B be reviewed and gave clear guidance, in particular to the spread of fire over the external envelope of the building. It also advised government to encourage consideration of the retrofitting of sprinklers in high-rise residential buildings. The government restricted changes to a certification scheme for window installers (ignoring the recommendation concerning entire external envelope of building). Regarding sprinklers, they simply reiterated advice following the Shirley Towers fire in Southampton. The reason given for not pushing sprinklers was a concern of being legally required to provide funding.
2013:	The Grenfell Tower project appeared £2m higher than previous estimate. Studio E suggests ACM panels as an alternative option. An April report reiterates project is over-budget and could fail. A month later, KCTMO and RBKC decides value for money to be key driver for project. In September Studio E meets cladding sub-contractor Harley Facades who recommend ACM panels. Exova produce final version of fire strategy containing no reference to cladding but says design will have 'no adverse effect' in relation to

	external fire spread but that this will be confirmed in future analysis. Studio E prepare specs for job. Celotex is listed as insulation material. Solid zinc and aluminium panels are listed for the cladding but with Reynobond ACM as one of three alternatives.
2014:	Government was advised by an All-Party Parliamentary Group to take into account new research, abandon Class 0 and also that sprinklers were now more cost-effective. These could all be dealt with by 'simple amendments' before a full review scheduled for 2016/17. The government refused. Dr Webb, expert advisor to committee, wrote warning of 'a major fire tragedy with loss of life'.[27]
2014:	Celotex insulation product, later used on Grenfell, failed a test at BRE with cement-fibre cladding panel. In a second test temperature monitors were reinforced with fire-resistant boards. This passed and allowed Celotex to claim the product was suitable for buildings over eighteen metres. They obtained certificate from LABC.[28]
2014:	Kinspan secured a test pass, but it was on a different material. Kingspan sought advice from Dr Barbara Lane (later to be expert witness to inquiry). She told the National House Building Council that she was 'deeply concerned … the use of highly combustible materials in residential buildings is now simply an accident waiting to happen'.[29]
2014:	A senior civil servant e-mailed a senior figure at the National House Building Council (NHBC) warning of potential dangers of plastic insulation used in cladding panels.[30]
2014:	NHBC state that if no test information was available a desktop study report from an expert would suffice.[31]
2014:	At Grenfell three contractors submit tender bids. Rydon at £9.249 million are lowest. Ahead of formal award, KCTMO meets with Rydon, despite legal advice that conversations of this nature were prohibited before the contract was awarded, to discuss reducing costs to £8.5 million. Rydon is successful contractor to design and build refurbishment. Harley became the cladding sub-contractor. and was Studio E retained. The design team requested permission from RBKC planning department to change from zinc to ACM panels. The certificate from BBA suggesting panels have Class 0 classification is seen. Planner approves switch and insists on cassette-type system. Mock-up of cladding system installed but

uses panels with fire-retardant core. Eventual cladding had far more combustible polyethylene.

Architect at Studio E e-mails Exova with drawings from Harley asking for guidance on correct positioning on cavity barriers. Advice that 'if the insulation … is combustible you will need to provide cavity barrier as shown on your drawing' is wrong as combustible insulation should not be used.

The Project manager at the TMO for Grenfell recalled the Lakanal House fire and requested asking for clarification regarding 'flame retardant requirement'. There is no evidence of a response or follow up and this was described in the inquiry as the 'last chance to avert disaster'.[32] Building control officer e-mails Studio E giving full plans approval 'subject to conditions'. No record of these conditions has been found.

2014: LFB issue deficiency notice in March ordering TMO to fix smoke-control system by May. Work delayed for another year.[33] By March the TMO had a backlog of 1,400 incomplete actions and 650 blocks requiring risk assessments.[34]

2014: Generic Risk Assessment 3.2 in national guidance for fire services revised. Stated that fire authorities should have contingency plans including an operational plan should stay-put become untenable.[35] However, no training for such a contingency was provided.

2014: Boris Johnson as Mayor of London cut ten fire stations, twenty-seven fire engines and 552 firefighters.

2015: NHBC write to Kingspan warning that K15 is to be rejected for use on high-rise blocks. Kingspan respond with solicitor's letter warning of legal action. NHBC moderates position and allows for desktop study.[36]

2015: Cladding fire, Torch Tower Dubai.

2015: Harley Facades purchases Celotex insulation from supplier SIG for Grenfell Tower. The building inspector raised question of cavity barriers. This was considered unnecessary by design team as 'ACM will be gone rather quickly in a fire!'.[37] Siderise, sub-contractor of cavity barriers e-mails Harley Facades design drawings highlighting a 'weak link for fire' at the top of the windows. The e-mail is not forwarded. Elsewhere, e-mails show combustible Celotex insulation has been selected to fill gaps around windows, despite non-combustible Rockwool being specified. Harley Facades orders Kingspan K15 insulation to supplement low availability of Celotex. KCTMO are not informed of switch. Rydon writes concerning 'poorly performing site'. Later inspections fail to spot flaws, including badly fitted or missing cavity barriers. Building inspector who signed off Grenfell relied on the BBA certificate to judge cladding was compliant.[38]

2015: After a fire at nearby Adair Tower, the LFB issued TMO with 'deficiency notice' demanding all doors have self-closers, followed by an enforcement notice.[39]

2016: NHBC publishes guidance setting out systems that were acceptable even without a desktop study. This included Kingspan's K15 and Celotex's RS5000 and even ACM panels so long as they had a Class B rating.[40]

2016: Cladding fire in Middle East. In 2016, the year the refurbishment at Grenfell was completed, a civil servant in the department responsible for building regulations admitted in writing that ACM panels were flammable when the polyethylene core is exposed.[41] A cladding supplier also wrote to the government department with 'grave concern' regarding ACM, combustible foam thermal-insulation boards and the number of buildings involved in UK. The response insisted the rules regarding flammable material on exterior of high-rise buildings are not ambiguous and 'if designer and building control body choose to use them it 'is up to them'.[42]

2016: A group representing the cladding industry agree Approved Document B required clarifying regarding materials used on exterior of buildings. An all-Party Parliamentary Group on Fire Safety continued to write to government with concerns. Dr Webb raised the matter with a senior civil servant responsible warning of a death toll ten to twelve times that of Lakanal (i.e. sixty-seventy-two people). Despite all this, a government report concluded there was no need for change.[43]

2016: Arconic warns French team of significant differences between PE and FR panels and required FR to be used in future. A similar warning was not given to UK.[44]

2016: LFB carried out inspection of Grenfell Tower and issued deficiency notice regarding fire doors not being self-closing. Investigations

	after the fire found forty-three of 129 doors had no self-closers and thirty-four that did failed to work properly.⁴⁵
2016:	In May RBKC building control issues completion certificate for project and KTCMO press release celebrates completion of refurbishment.
2017:	In April, LFB Commissioner Dany Cotton, wrote to government ministers warning of blocks with 'significant compartmental issues'. The RBKC sent the letter from the LFB to the TMO who forwarded it to consultant risk assessor. They believed they did not have such cladding. He replied that the cladding complied with building regulations.⁴⁶

On 13 June, the day before the fire, and seen by the civil servant responsible, the Fire Sector Federation reported that Approved Document B was 'out of date, placing businesses and communities all over the UK at potentially fatal risk'.⁴⁷ Also, the day before the fire the TMO at RBKC had 287 actions outstanding (128 more than a year old).⁴⁸

The Grenfell Inquiry Phase Two Report

As I write this final note in September 2024, over seven years after the fire, the Grenfell Tower Inquiry Phase Two Report has just been published. It is scathing of the relevant parties involved. It concluded, 'the fire at Grenfell Tower was the culmination of decades of failure by central government and other bodies in positions of responsibility in the construction industry'.

It found the Department for Communities and Local Government had received numerous warnings about the risks between 2012 and 2017. Indeed, they were well aware since the 1991 fire at Knowsley Heights. By 2013 they knew that Approved Document B was unclear and not properly understood and by February 2016 they knew that panels were routinely being used on high-rise buildings in breach of guidelines.

Regarding the British Research Establishment, the report concluded: 'much of the work it carried out in relation to testing the fire safety of external walls was marred by unprofessional conduct, inadequate practices, a lack of effective oversight, poor reporting and a lack of scientific rigour'. This led to the risk of 'manipulation by unscrupulous product manufacturers, as happened.'

The companies involved in making and selling the rain screen cladding panels and insulation products were described as engaging in 'systematical dishonesty' with 'deliberate and sustained strategies to manipulate the testing processes, misrepresent test data and mislead the market'.

Arconic Architectural Products manufactured and sold the Reynobond 55 PE rain screen panels used in the external wall of Grenfell Tower. These were the ACM product consisting of two thin sheets of aluminium with a polyethylene core. Arconic 'concealed important information from the BBA, in particular the test data relating to the product in cassette form, that showed that it performed much worse than in rivetted form. It caused the BBA to make statements in the certificate that Arconic knew to be false and misleading'.

Celotex manufactured RS5000, the combustible polyisocyanurate foam insulation. They 'embarked on a dishonest scheme to mislead its customers and the wider market'. From 2011 they sold it as Class 0 fire performance which they knew was 'false and misleading'. They presented RS5000 as suitable and safe for use on Grenfell Tower despite knowing it was not.

Kingspan arguably received even stronger criticism. They 'knowingly created a false market in insulation for use on buildings over 18 metres in height by claiming that K15 had been part of a system successfully tested under BS 8414 and could therefore be used in the external wall of any building over 18 metres in height regardless of its design or other components.' The inquiry found this claim was false and Kingspan 'knew K15 could not honestly be sold as suitable for use in the external walls of buildings over 18 metres in height generally'.

Further they concealed from the British Board of Agrément, BBA, the fact that the product it was selling, to which the 2008 certificate referred, was not the same as the system tested in 2005. The certificate also contained three important statements about the fire performance of K15 that were untrue. In 2009, Kingspan also obtained a certificate, this time from the LABC, that contained false statements about its product relating to high-rise buildings. Kingspan then 'dishonestly relied' on the results of those tests to support the sale of K15.

The British Board of Agrément certified the compliance of products with the requirements of legislation. The inquiry found that they were susceptible to the 'dishonest strategies of Arconic and Kingspan' due to the following reasons: their incompetence; a failure to adhere robustly to the system of checks it had put in place; and, lastly, an 'ingrained willingness' to put customers wishes ahead of high standards. Their own commercial interests in attracting and retaining customers influenced their requirement for rigour and independence.

Local Authority Building Control (LABC) and the National House Building Council (NHBC) also received criticism. The former had 'a complete failure ... over a number of years to take basic steps to ensure that the certificates it issued in respect of them were technically accurate'. The latter were 'unwilling to upset its own customers and the wider construction industry by revealing the scale of the use of combustible insulation in the external walls of high-rise buildings, contrary to the statutory guidance.' Importantly, the inquiry found 'that the conflict between the regulatory function of building control and the pressures of commercial interests prevents a system of that kind from effectively serving the public interest'.

All in all, a systemic failure of government and institutions responsible for the safety of buildings, creating a situation of which unscrupulous companies took full advantage. The companies responsible for the Grenfell refurbishment also received criticism. Studio E, Rydon and Harley all took 'a casual approach' to contractual relations as well as to their obligations and responsibilities relating to fire safety. Everyone involved in the choice of materials and design of the external wall failed to act in accordance with the standards of a reasonably competent person in their position. It was found they were unfamiliar, or did not understand, the relevant provisions of the Building Regulations, Approved Document B or industry guidance. A cavalier and casual attitude persisted in regard to fire safety when they should have been well aware of the risks of using combustible materials in the external walls of high-rise buildings.

The London Fire Brigade was blamed for a 'chronic lack of effective management and leadership, combined with an undue emphasis on process'. Problems deemed 'undeserving of change or too difficult to resolve' were ignored even when they concerned operational or public safety. There was a failure to respond quickly to issues and identify training needs. This I certainly agree with.

It seems to me the problem after Lakanal was that they acknowledged that evacuation needed to be considered but the very process of a quick evacuation inevitably meant changing some procedures quite radically. This the Brigade was loath to do. Many of these procedures had developed after serious fires in the past, some fatal, resulting in coroner's recommendations.

What they seem to find difficult is to balance both sides of the risk/reward equation and to assess different risks together. That is why we get working on, or near, water procedures which inevitably result in the press reporting an incident where the emergency services stand by as a man is face-down in a pond while they wait for a specialist Water Rescue team. We address one risk and create another. They stop snatch rescues and pat themselves on the back for reducing risks to firefighters. But at the same time ignore the rescues that may have occurred had firefighters accepted a specific level of risk in a particular situation.

A similar line of thinking resulted in the firefighter's lift being scrapped and the lowering of strength standards. They focus on their perceived benefit around recruitment targets and utterly ignore the risks to the public from having firefighters who cannot rescue an adult casualty on their own.

In the same way, high-rise procedure had developed for very good reasons. However, the only way to evacuate a high-rise block quickly with one staircase and no alarm is to send someone above the fire to knock on doors. And this can't be done quickly if we insist they've all got to have started up in breathing apparatus and carrying a charged length of hose. Of course this goes against established procedures. But the answer isn't to ignore it, pretend it won't happen or hope firefighters will break procedures and then not discipline them in a high-profile incident *but* do so when the media isn't around.

When I first joined the fire brigade we were taught the following: we risk a lot to save saveable life, we risk a little to save saveable property, we don't take a risk to save lives or property that have already been lost. We seem to have forgotten this simple rule of thumb, as well as what we are here for.

There are, however, a few points I disagree with concerning the inquiry. I have already stated that I believe any window of opportunity for evacuation was a little earlier and narrower than the inquiry found. Additionally, I feel the results of such a decision are far less certain either at Grenfell or, importantly, at some future incident.

The report acknowledged that 'firefighters must be allowed to exercise discretion in how best to carry out their instructions'. However, it went on to note the number of times 'crews were despatched to the highest floors of the tower in response to calls for assistance failed to reach their destinations because they decided to help people they encountered on the stairs on their way up'. They recommended that the National Fire Chiefs' Council considers 'whether, and if so in what circumstances, firefighters should be discouraged from departing from their instructions on their own initiative and provide appropriate training in how to respond to a situation of that kind'.

The implication is that crews made the wrong decision. But let's look at the practicalities. You make your way up a staircase full of hot thick smoke to perform a rescue. Halfway up a mother and two children stumble into you. What exactly do you do? If you know they are unlikely to make it down the stairs you can hardly leave them. You have the choice of making a rescue or leaving them and attempting a possible rescue further on. We can rewrite procedures and allow crews to split up but, having imposed the rule of never leaving your crew and going off on your own on pain of discipline and dismissal, you can hardly blame firefighters for sticking to it.

Having said that in a situation such as Grenfell I wouldn't blame a crew for doing just that: one firefighter leading them to safety and returning after. But we recall a crew had just that situation in the early stages and led them to safety with one resident collapsing on the way down, requiring both firefighters to get them all out.

Let us imagine we follow the inquiry's implication and leave the residents on the stairs to make their own way and continue to the original rescue. Even if you are successful and find those casualties, on your way down you may find the stairs blocked by the residents you left on the stairs, now collapsed from smoke inhalation.

The most sensible solution is to rescue the ones in front of you and radio back for another crew to take over your original task. I'm not sure how one can justify the alternative. Of course, there are no easy solutions and risks either way. But it does highlight to me that these sorts of comments don't fully appreciate the practicalities of their suggestions.

Another area where I felt the report didn't appreciate practicalities was in familiarisation visits, especially to high-rise blocks. It referenced some 8,500 in London. This equates to about 80 per station. Some stations, such as my last station at Addington in South London, had fewer than ten such blocks. Inner city stations might have 200. If we add all the other types of premises, visits, community fire safety work, statutory fire safety work, hydrants, training and other work, it is simply not possible for some stations to do the range and depth of inspections the inquiry asked for.

You could hire people for each borough to conduct these visits. You could drastically increase the number of fire safety officers. But if the plan is to get station staff to do all these visits it is, in my opinion, doomed to failure. They are not fire safety officers or building control inspectors. They do not have the time or expertise to draw up the kind of detailed plans for every building deemed a risk on their ground any more than the local police station has the resources to check every single car on their ground.

Having said all that, in general I was very impressed with the inquiry, the scope, detail and professionalism of all involved. I do think it was the wrong way round and could and should have been completed much sooner. Nor should it have prevented government acting decisively back in 2017. A system of works to remove all the offending panels should have been completed within a year. That we still have these problems seven years later is a scandal upon a scandal.

It would have been possible to have made a wider political point in this book. Anyone working in the police, NHS, education, social services, prisons and many other areas of the public sector will have experienced a similar dumbing down of standards and drift away from quality. Yet it is sometimes difficult to see the bigger picture. So I decided to confine the book to issues around building regulations and fire safety. Hopefully I have demonstrated Grenfell should be seen as a systemic failure across the board over several decades rather than isolated to one block, council or government.

I finish this three months after Labour came into power after the 2024 General Election. Time will tell if they perform any better than their predecessors. But our problems are not simply confined to one fire, as tragic as it was. It's not even confined to fire safety or building regulations. It's layers upon layers of systemic failures across the board over many decades. At the heart is forgetting what we here for or why we are doing things. Whether it's a government, local authority, building control or Fire Brigade prioritising appearance over substance or processes over outcomes, it's about forgetting what quality means in terms of a service and losing the integrity and confidence to maintain that level. As well as losing the will to put the effort in or pay for that quality, or stand up to forces that degrade that quality, whether deliberately or indirectly.

With that in mind I will finish this with what I'd like to see by the end of this particular government:

Building Regulations and Fire Safety legislation that is fit for purpose.

A Fire Brigade that prioritises public safety in every policy, procedure or process.

The cladding scandal resolved with no building having flammable cladding and no flat owner financially out of pocket.

All those affected by the Grenfell fire adequately supported.

Building Control and Fire Safety adequately funded with enough officers and powers to ensure quality of buildings and fire safety.

Institutions such as the British Research Establishment, Local Authority Building Control and The National House Building Council developing a reputation for independence and integrity.

Government departments that place the interests of the public first.

Finally, a government that we can trust.

All this is perfectly possible. If none of it happens, it's because there's no political will to make it happen and probably a fair amount to prevent it. And we should not forget this when we next come to use our vote.

Notes

Introduction
1. https://www.bbc.co.uk/news/av/uk-40332427

Part I
1. Apps, *Show me The Bodies*, pp.32-3
2. Prosser and Taylor, *The Grenfell Tower Fire*, pp.181-2
3. Apps, op. cit., p.35
4. Ibid., p.37
5. Ibid., p.38
6. Prosser and Taylor, op. cit., pp.122-7
7. https://www.fbu.org.uk/news/2022/05/13/fbu-calls-grenfell-building-safety-body-be-nationalised
8. Apps, op. cit.,p.39
9. Ibid., p.42
10. Ibid., p.43
11. Prosser and Taylor, op. cit., p.120
12. Ibid., p.231

Part II
1. Apps, op. cit., p.106 Ibid., p.127
2. Ibid., p.44
3. Ibid., p.133
4. Ibid., pp.45-6
5. Ibid., pp.108-9
6. Ibid., p.110
7. Ibid., pp.133-4
8. Ibid., p. 135
9. Ibid., p. 135
10. Ibid., p.137
11. Prosser and Taylor, p.235
12. Ibid. p.236
13. Ibid., p 239
14. Apps, op. cit., pp. 69-70
15. Ibid., p.71
16. Prosser and Taylor, op. cit., p.241
17. LFB PN 633 *High-Rise Firefighting*
18. https://assets.grenfelltowerinquiry.org.uk/documents/LFB%20Policy%2C%20High-rise%20firefighting%2C%20issued%2026%3A11%3A2008%20and%20%28policy%20633%2C%20reviewed%2001%3A06%3A2015%29.%20LFB00001256.pdf

19. Apps, op. cit., p.233
20. Ibid., pp.64-5
21. Ibid., p.268
22. Ibid., p.78
23. Ibid., p.205
24. Ibid., p.111
25. Ibid., pp.115-16
26. Ibid., p.113
27. Prosser and Taylor, op. cit., p.61
28. Ibid., p.78
29. Ibid., p.82
30. Ibid., p.81
31. Ibid., p.85
32. Ibid., p.88
33. Ibid., p.91
34. Ibid., p.95
35. Apps, op. cit., pp.68-70
36. Ibid., p.74
37. Ibid., pp.139-41
38. Ibid., pp.143-4
39. Ibid., p.142
40. Ibid., p. 144
41. Ibid., p.180
42. Ibid., p. 205
43. Ibid., p. 234
44. Bisby, expert witness report, 2018: 21
45. Lane, Barbara, expert witness report, Grenfell Inquiry: 51
46. Prosser and Taylor, op. cit., 130
47. Apps, Peter, 2020: Inside Housing Magazine
48. Apps, op. cit., p. 145
49. Ibid., p. 211-2
50. Ibid., p. 76
51. Ibid., p. 77
52. Ibid., pp. 78-81
53. Ibid., pp. 116-7
54. Ibid., p. 213
55. Ibid., pp. 214-5
56. Ibid., p. 237
57. https://www.ifsecglobal.com/wp-content/uploads/2016/07/The-Fire-Performance-of-Building-Envelopes-by-Steven-Howard-BRE-Global.pdf
58. Apps, op. cit., p. 83
59. Ibid., p. 235

Part III

1. Inquiry Phase One Report, Executive Summary 2.19 b
2. Apps, op. cit., p. 11
3. Ibid., p. 11
4. Lane, op. cit., L.6

5. Apps, Peter, Grenfell Tower Inquiry diary week 44, *Inside Housing*, 22.07.21
6. https://www.insidehousing.co.uk/insight/grenfell-tower-inquiry-diary-week-44-ive-never-seen-a-fully-compliant-firefighting-lift-in-any-local-authority-building-to-this-day-actually-71813
7. Lane, op. cit., p. 9.7
8. Ibid., p. 10.10
9. https://assets.grenfelltowerinquiry.org.uk/documents/Dr%20Barbara%20Lane%20report%20-%20section%2010_0.pdf
10. Lane, op. cit., p. 2.11.24
11. Ibid., 2.11.35
12. Apps, 5.07.21, Inside Housing Magazine
13. Lane, op. cit., p. 2.13.14-5
14. Ibid., p. 2.14.5
15. Prosser and Taylor, op. cit., p. 261
16. Ibid., p.249
17. Ibid., p.273
18. Lane, op. cit., p. 14.8.18
19. Prosser and Taylor, op. cit., p.273
20. Ibid., pp. 274-5
21. Ibid., p. 275
22. Ibid., p. 277
23. Lane, op. cit., p. 19.6.61
24. Prosser and Taylor, op. cit., p. 283
25. https://www.youtube.com/watch?v=Mc4JI8caQPU
26. https://www.dailymail.co.uk/video/news/video-1484865/Some-Grenfell-Tower-disaster-victims-never-identified.html
27. https://www.youtube.com/watch?v=Gek20riA9l8
28. https://assets.grenfelltowerinquiry.org.uk/KHA00000001_Witness_Statement_of_Karl_Harrison_dated_08.06.2022.pdf
29. https://www.independent.co.uk/voices/grenfell-tower-fire-one-year-one-kensington-a8397276.html
30. Apps, op. cit., p. 296
31. Apps, 2022: 263
32. Lane, op. cit., p. 20.6.6
33. Ibid., p. 18.9.3
34. Apps, op. cit., p. 265
35. Ibid., p. 237
36. Ibid., p.217-8
37. https://stevedude68.com/2019/06/22/grenfell-tower-part-one/
38. https://stevedude68.com/2019/09/02/grenfell-tower-part-2-stay-put/
39. https://stevedude68.com/2019/10/29/grenfell-tower-systematic-failure/
40. Apps, op. cit., p. 207
41. Ibid., p. 298
42. Ibid., p. 249
43. Ibid., pp. 253-4
44. Lane, op. cit., p. 20.4.21
45. Apps, op. cit., p. 258
46. Ibid., p. 247

47. Ibid., p. 266
48. Ibid., p. 250
49. Ibid., p. 254
50. Ibid., p. 252
51. Lane, op. cit., p. 11.21.15
52. Prosser and Taylor, op. cit., 249
53. Apps, op. cit., p. 294
54. Prosser and Taylor, op. cit., 249
55. https://www.fbu.org.uk/news/2023/06/14/grenfell-tower-fire-crime-caused-deregulation-says-fbu
56. Apps, op. cit., p. 145-6
57. Ibid., p. 45-6
58. Ibid., p. 265
59. https://www.fbu.org.uk/publications/high-rise-buildings-breathing-apparatus-safety-campaign
60. https://www.bbc.co.uk/news/articles/ce7kyp4z4lxo

Appendix
1. Apps, op. cit., pp.32-3
2. Ibid., p. 37
3. Ibid., p. 38
4. Ibid., p. 39
5. Ibid., p. 42
6. Ibid., p. 43
7. Ibid., p. 106
8. Ibid., p. 127
9. Ibid., p. 44
10. Ibid., p. 133
11. Ibid., pp. 45-6
12. Ibid., pp. 108-9
13. Ibid., p. 110
14. Ibid., pp. 133-4
15. Ibid., p. 135
16. Ibid., p. 135
17. Ibid., p. 137
18. Ibid., p. 61
19. Ibid., p. 233
20. Ibid., pp. 64-5
21. Ibid., p. 178
22. Ibid., p. 268
23. Ibid., p. 205
24. Ibid., p. 111
25. Ibid., pp. 115-16
26. Ibid., p. 113
27. Ibid., p. 74
28. Ibid., pp. 139-41
29. Ibid., pp. 143-4
30. Ibid., p. 142

31. Ibid., p. 144
32. Ibid., p. 180
33. Ibid., p. 205
34. Ibid., p. 234
35. Ibid., p. 256
36. Ibid., pp. 144-5
37. Ibid., p. 181
38. Ibid., p. 116
39. Ibid., pp. 211-12
40. Ibid., p. 145
41. Ibid., p. 76
42. Ibid., p. 77
43. Ibid., p. 78-81
44. Ibid., p. 116-17
45. Ibid., p. 213
46. Ibid., p. 237
47. Ibid., p. 83
48. Ibid., p. 235

Bibliography

Apps, Peter, *Show Me the Bodies: How We Let Grenfell Happen* (Oneworld Publications, London, 2022).
——, *The building safety crisis: far from over*, Inside Housing, 29.03.23, https://www.insidehousing.co.uk/insight/the-building-safety-crisis-far-from-over-80725
——, *Grenfell Tower refurbishment: a timeline*, Inside Housing 27.11.2020, https://www.insidehousing.co.uk/insight/grenfell-tower-refurbishment-a-timeline-68533
——, *Grenfell smoke control system 'did not comply' with guidance and was untested on other buildings*, Inside Housing, 05.07.21, https://www.insidehousing.co.uk/news/grenfell-smoke-control-system-did-not-comply-with-guidance-and-was-untested-on-other-buildings-71443
——, *Grenfell Tower Inquiry diary week 44*, Inside Housing, 22.07.21, https://www.insidehousing.co.uk/insight/grenfell-tower-inquiry-diary-week-44-ive-never-seen-a-fully-compliant-firefighting-lift-in-any-local-authority-building-to-this-day-actually-71813
Grenfell Tower Inquiry: Phase One Report, https://www.grenfelltowerinquiry.org.uk/phase-1-report
Grenfell Tower Inquiry: Phase Two Report Overview, https://www.grenfelltowerinquiry.org.uk/sites/default/files/CCS0923434692-004_GTI%20Phase%202_Report%20Overview_E-Laying_0.pdf
Grenfell Tower Inquiry: Phase Two Report Volume 5, Part 8, https://www.grenfelltowerinquiry.org.uk/sites/default/files CCS0923434692-004_GTI%20Phase%202%20Volume%205_BOOKMARKED.pdf
Kernick, Gill, *Catastrophe and Systemic Change, Learning from the Grenfell Tower Fire and Other Disasters* (London Publishing Partnership, London, 2021).
Prosser, Tony, and Mark Taylor, *The Grenfell Tower fire: Benign neglect and the road to an avoidable tragedy* (Pavilion, Shoreham by Sea, 2020).

Dear Reader,

We hope you have enjoyed this book, but why not share your views on social media? You can also follow our pages to see more about our other products: facebook.com/penandswordbooks or follow us on X @penswordbooks

You can also view our products at www.pen-and-sword.co.uk (UK and ROW) or www.penandswordbooks.com (North America).

To keep up to date with our latest releases and online catalogues, please sign up to our newsletter at: www.pen-and-sword.co.uk/newsletter

If you would like a printed catalogue with our latest books, then please email: enquiries@pen-and-sword.co.uk or telephone: 01226 734555 (UK and ROW) or email: uspen-and-sword@casematepublishers.com or telephone: (610) 853-9131 (North America).

We respect your privacy and we will only use personal information to send you information about our products.

Thank you!